PERSONNEL TRAINING

PTP

PUBLICATIONS

Second Edition

AF887240

ET
CLASSROOM TRAINING BOOK

ASNT

Second Edition
first printing 04/18
second printing 08/21
third printing 01/25
ebook 04/18

Errata, if available for this printing, may be obtained from the product page on ASNT's website, source.asnt.org. Errata for out-of-print books may be obtained by contacting ASNT Customer Support. Ebooks contain all corrections and updates, including the latest errata.

ISBN: 978-1-57117-417-8 (print)
ISBN: 978-1-57117-418-5 (ebook)

Printed in the United States of America

Published by:
The American Society for Nondestructive Testing Inc.
1201 Dublin Road, Suite #G4
Columbus, OH 43215
asnt.org

Edited by: Toni Kervina, Educational Materials Editor
Bob Conklin, Instructional Designer
Assisted by: Cynthia M. Leeman, Educational Materials Supervisor
Layout and Illustrations by: Synthia Jester, Graphic Designer
Joy Grimm, Production Manager

Tim Jones, Senior Manager of Publications

ASNT Mission Statement:
ASNT's mission is to advance the field of nondestructive testing.

ASNT Code of Ethics:
The ASNT Code of Ethics was developed to provide members of the Society with broad ethical statements to guide their professional lives. In spirit and in word, each ASNT member is responsible for knowing and adhering to the values and standards set forth in the Society's Code. More information, as well as the complete version of the Code of Ethics, can be found on ASNT's website, asnt.org.

Acknowledgments

The American Society for Nondestructive Testing, Inc. is grateful for the technical expertise, knowledge, and contributions of James E. Cox, Hussein M.A. Sadek, and the entire team of electromagnetic testing method reviewers, contributors, and subject matter experts who tirelessly worked to develop the second edition of the *Electromagnetic Testing Classroom Book.* The second edition builds on the first edition, originally written by Hussein M.A. Sadek in 2006. The following individuals assisted with this edition of the *Electromagnetic Testing Student and Instructor Packages*, which include the *Classroom Training Book* and *Student/Lecture Guides*:

James E. Cox, Technical Editor – JECNDT, LLC
Michael C. Brown – Metal-Matic, Inc.
John Chen – Schlumberger
Sebastian Fernandes – Petrofac
Dinesh Gupta – Satyakiran Engineers Pvt., Ltd.
Navita Gupta – Satyakiran School of NDT
Gerard Hacker – Teledyne Brown Engineering
William G. Hoffmann – Magnetic Analysis Corp.
Mark A. Johnson – Johnson NDT
Bharath Kumar Kodumuru – Textron India Pvt., Ltd.
Dr. Martin C. Lugg – TSC Inspection Systems
Joseph L. Mackin – International Pipe Inspectors Association, Inc.
John A. Markanich – Perryman Company
Steve A. Micich – AAR Corp.
Nirav Dave – Ratnamani Metals & Tubes Ltd.
Hussein M.A. Sadek – Technologies Consulting International, Inc.
Dharmveer Singh – GE
Dr. Michael C. Smith – TSC Inspection Systems
Dr. Roderic K. Stanley – NDE Information Consultants
Abhishek Wakaskar – Triumph Group

The Publications Review Committee includes:

Joseph L. Mackin, Chair – International Pipe Inspectors Association
Martin T. Anderson – Alaska Technical Training
Mark R. Pompe – West Penn Testing Group

Foreword

Purpose

The American Society for Nondestructive Testing, Inc. (ASNT) has prepared this series of Personnel Training Publications to provide an overview in a classroom setting of a given nondestructive testing method. Each classroom training book in the series is organized to follow the body of knowledge found in *ANSI/ASNT CP-105: ASNT Standard Topical Outlines for Qualification of Nondestructive Testing Personnel.* Level I and Level II candidates should use this classroom training book as a preparation tool for nondestructive testing certification. Note, however, that an NDT Level I or Level II may be expected to know additional information based on industry or employer requirements.

Supplementary Material

Although the classroom training book may be purchased and read as a standalone product, it is intended to be used in conjunction with the Lecture Guide and PowerPoint™ presentation for instructors and Student Guide for students. These guides contain a condensed version of the material in the classroom training book and quiz questions per chapter (lesson) for review purposes.

Contents

LEVEL >I

1

Introduction to Electromagnetic Testing

Early Observations of Magnetic Events

Electromagnetic testing (ET) is one of the oldest nondestructive testing (NDT) methods. Thales of Miletus (600 BC) first recorded that rubbing amber induced a state in which the amber would attract other light objects. The Greek word for amber is electron. Thales also mentioned the remarkable powers of lodestone (iron oxide), also known as magnetite.

Democritus (400 BC) provided concepts of an atomic structure of matter. Democritus' fifth principle states that "varieties of all things depend on the varieties of their atoms, in number, size, and aggregation." Many electromagnetic tests are used to identify the composition of specific material and the discontinuities that occur in those materials.

By 1200 AD, the use of the magnetic compass was reported in China. At about the same time, Englishman Alexander Neckam also reported the use of the compass in navigation. In the year 1600, the physician William Gilbert wrote a comprehensive description of his 18 years of experiments and his theory of magnetism in the book *De Magnete*.

Örsted's Discovery

Hans Christian Örsted, shown in Figure 1a, observed that a wire connecting the ends of a voltaic battery affected a compass in its vicinity. In his published account in 1820, he explained that the current itself was

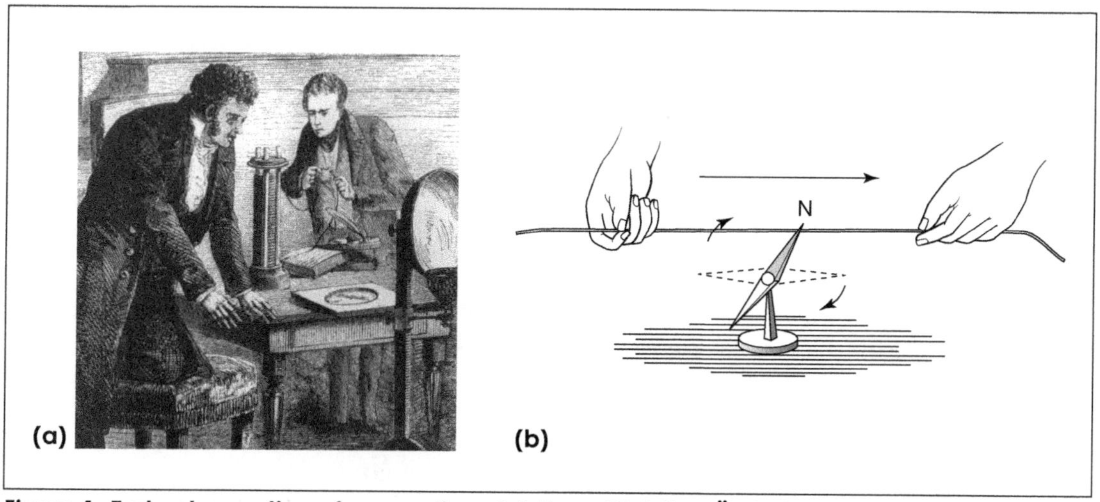

Figure 1. Early observation of magnetism: (a) Hans Christian Örsted with student; and (b) influence of electric current on compass needle.

the cause of the action, and that the electric conflict acts in a revolving manner, that is, a compass placed near a wire transmitting an electric current tends to set itself perpendicular to the wire and always points toward the wire as the compass is moved around the wire. The space in which these forces act may therefore be considered a magnetic field, as shown in Figure 1b. Örsted's discovery meant that the lines of magnetic force are at right angles to the wire and are therefore circular and perpendicular to the wire.

Faraday's Law of Electromagnetic Induction

Electromagnetic testing originated when Michael Faraday discovered the effect of electromagnetic induction while experimenting with coils of wire and a battery in 1831. Joseph Henry independently discovered the same effect the following year, but Faraday was the first to publish the results of his experiments. He noticed that by connecting one coil to a battery, there was an instant electrical current through a second coil placed near the first coil when he switched the battery on and off, as diagrammed in Figure 2. He concluded that the second current was in the opposite direction of the first current.

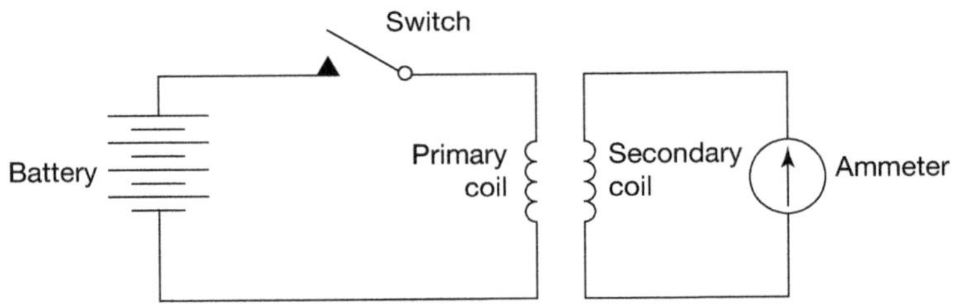

Figure 2. Electromagnetic induction in a direct current circuit.

Similarly, Faraday found that moving the secondary circuit toward the primary induced a current opposite to the primary current in the primary coil. Also, moving the secondary circuit away from the primary induced a current in the same direction as the primary current in the primary coil. It was James Clerk Maxwell who explained that "the direction of the secondary current is such that the mechanical action between the two conductors is opposite to the direction of motion, being a repulsion when wires are approaching, and an attraction when they are receding." This electromotive force was observed by Faraday but was given more systematic treatment by Heinrich Lenz.

Development of Induced Currents

Electromagnetic induction had not been observed nor explained before the 19th century. James Clerk Maxwell, shown in Figure 3, summarized the first 50 years of electromagnetism in the book *A Treatise on*

Figure 3. James Clerk Maxwell.

Electricity and Magnetism in 1873. Maxwell explained that conjectures had been made as to the relationship between magnetism and electricity, but the laws of these relationships had remained entirely unknown.

David E. Hughes performed the first recorded electromagnetic test in 1879. He was able to distinguish between different metals by observing and noting a change in excitation frequency resulting from effects of a test material's resistivity and magnetic permeability.

James Clerk Maxwell

James Clerk Maxwell conceived and published the comprehensive group of relations to the electromagnetic field known as Maxwell's equations, which offered a mathematical representation of the principles of electromagnetic testing. Maxwell's remarkable achievement of integrating the available knowledge concerning electromagnetic circuits and fields provides the basis for analysis of all basic eddy current and electromagnetic induction problems and most current electromagnetic theory.

Several ideas for electromagnetic methods of detection emerged in the 1930s. Most of these relied on inductive measurements of the leakage field. This required either a high-frequency magnetizing current, which limited the technology to near-surface discontinuity detection, or coils moving at constant velocity over the surface.

In 1946, the first practical system for the automatic, electromagnetic measurement of flux leakage fields was designed by C.H. Hastings.

Hastings demonstrated that he could detect both surface and subsurface discontinuities near the bore surfaces of steel tubes. He also noted an empirical correlation between the amplitude of leakage signals and the depths of surface cracks.

Friedrich Förster

The first eddy current instrument was developed in 1926. In 1950, Friedrich Förster developed the first instruments with impedance-plane signal displays, which were used to discriminate between different material parameters. Förster's introduction of quantitative test equipment and practical techniques for analysis of test signals on the impedance plane contributed to the rapid development and acceptance of electromagnetic tests from 1950 to 1965. Since 1965, more advanced electromagnetic testing applications have been introduced and today are used in nearly every technical industry.

Basic Principles of Electromagnetic Testing

ET is an NDT method that is based on the principles of electromagnetic induction. Electromagnetic simply means that electricity and magnetism are used. The flow of electricity, under certain circumstances, can cause magnetism. Magnetism, under certain circumstances, causes the flow of electricity. When an alternating current (AC) is passed through a coil, a changing magnetic field is generated. As the coil is placed near a conductive test object, the magnetic field induces current (eddy currents), as illustrated in Figure 4.

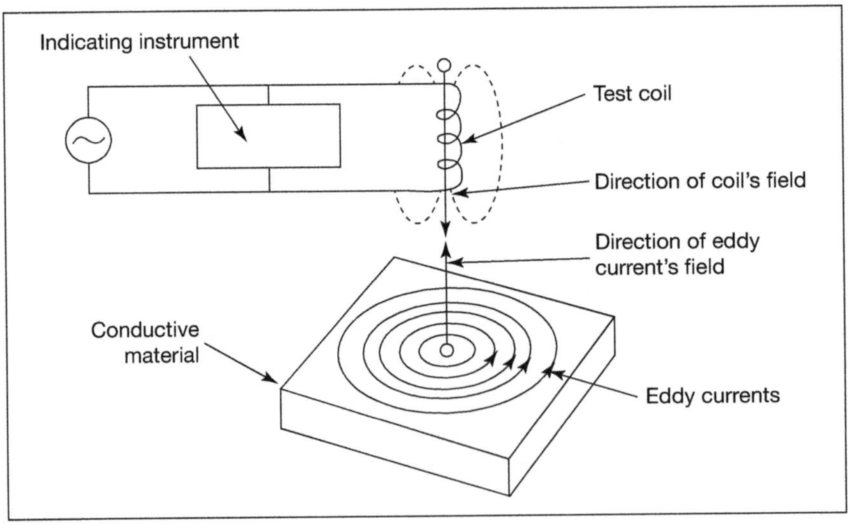

Figure 4. Basic eddy current test system.

The flow of eddy currents depends on the physical and electrical characteristics of the test object. As an eddy current flows in the test object, it causes a fluctuating magnetic field of its own. The magnetic field from the eddy currents is always in opposition to the coil's magnetic field, as also depicted in Figure 4. Thus, when the test coil is placed on a conductive material, the strength of the coil's magnetic field is lessened. This change in the magnetic field causes a change in the current flowing through the coil that, in turn, causes a change in the impedance of the coil. This change in impedance is detected by the readout display placed in the test circuit.

ET instruments measure and display these impedance coil changes and allow the technician to evaluate valuable information regarding the properties and condition of the test object.

ET, like some other nondestructive test methods, involves the application of electromagnetic energy to evaluate the condition of test objects. The energy interacts with the material, and the interaction process is analyzed to ascertain the condition of the material. Electromagnetic techniques include eddy current testing, alternating current field measurement, and remote field testing.

Although all electromagnetic techniques are governed by Maxwell's equations, the distinctive nature of each technique stems from differences in excitation frequencies, the nature of the sensors used, and the signal analysis techniques for characterizing the state of the test object. For example, eddy current inspections for near-surface indications are normally done in the kilohertz or even megahertz range. Remote field testing (RFT) is often performed in the hertz range. Each technique has acceptable discontinuity sensitivity when used correctly.

Principles of Magnetic Flux Leakage Testing

Magnetic flux leakage testing (MFL) is an NDT method that can provide a quick assessment of the integrity of ferromagnetic material. MFL may be further subdivided into direct current (DC) and alternating current (AC) techniques. Magnetic flux leakage uses DC in many systems, 60 Hz in AC yokes, or even higher frequencies for detecting very shallow surface cracks.

This method involves magnetization of the test object by a permanent magnet or by passing a direct current directly through a coil, creating an electromagnet. Internal conductors carrying DC or unidirectional short current pulses are often used for creating magnetic flux leakage in hollow ferromagnetic parts. The presence of a discontinuity on or near the surface of the test object disturbs the magnetic flux lines and results in a local leakage field around the discontinuity.

The magnetic flux leakage field can be detected using a variety of techniques including magnetic particles, noncontact sensors—such as a hall effect probe, or a simple induction coil. Hall elements oriented perpendicularly to the part surface are sensitive to the tangential leakage field. Induction coils may or may not be wound on a ferrite core and can or cannot be excited by AC. In the AC excitation case, they are called ferroprobes. Note: The application of magnetic particles to form indications is the primary technique of another NDT method, magnetic particle testing (MT), which is covered in a separate *Classroom Training Book.*

Personnel Qualification

It is imperative that personnel responsible for electromagnetic testing are trained and qualified with a technical understanding of the equipment and materials they will be using. They also need to have adequate test procedures for each task.

The American Society for Nondestructive Testing (ASNT) has published guidelines for training and qualifying NDT personnel since 1966. These are contained in *Recommended Practice No. SNT-TC-1A: Personnel Qualification and Certification in Nondestructive Testing. SNT-TC-1A* describes the knowledge and capabilities of NDT personnel in terms of certification levels.

A related document, *ANSI/ASNT CP-189: ASNT Standard for Qualification and Certification of Nondestructive Testing Personnel*, was approved by the ASNT Board of Directors in 1989. The intent was to produce a new document that provided strict requirements rather than simply guidelines and recommendations. ASNT obtained American National Standards Institute (ANSI) accreditation to process this document through a consensus balloting process that would recognize *CP-189* as a national standard. The first successful consensus document became *ANSI/ASNT CP-189* (1991).

Note that some industries may use different certification standards. For example, the aerospace industry uses National Aerospace Standard *NAS 410: NAS Certification & Qualification of Nondestructive Test Personnel* and *Airlines for America A4A SPEC 105, Guidelines for Training and Qualifying Personnel in Nondestructive Testing.* Both documents are similar in purpose and scope to ASNT guidelines and standards. Other training and qualification schemes are available such as Personnel Certification in Non-Destructive Testing (PCN) offered by the British Institute of Non-Destructive Testing (BINDT) in the United Kingdom and ASME Nondestructive Examination (ANDE) Personnel Certification by the American Society of Mechanical Engineers in the United States. Many countries around the world have tried to standardize their approach to NDT personnel qualification and certification. *ISO-9712* is another standard that is often cited.

Levels of Qualification

There are three basic levels of qualification applied to NDT personnel and used by companies that follow *SNT-TC-1A* or *CP-189*: Level I, Level II, and Level III.

An individual in the process of becoming qualified or certified to Level I is considered a trainee. A trainee does not independently conduct tests, interpret, evaluate, or report test results of any NDT method. A trainee works under the direct guidance of certified individuals.

Level I Personnel
Level I personnel are qualified to perform the following tasks:
- Perform specific standardizations and nondestructive tests in accordance with specific written instructions.
- Record test results. Normally, the Level I does not have the authority to sign off on the acceptance and completion of the nondestructive test unless specifically trained to do so with clearly written instructions.
- Perform NDT job activities in accordance with written instructions or direct supervision from Level II or Level III personnel.

Level II Personnel
Level II personnel are qualified to perform the following tasks:
- Set up and standardize equipment.
- Interpret and evaluate results with respect to applicable codes, standards, and specifications.
- Organize and report the results of nondestructive tests.
- Exercise assigned responsibility for on-the-job training and guidance of Level I and trainee personnel.
- Be thoroughly familiar with the scope and limitations of each method for which the individual is certified.

Level III Personnel
A Level III is responsible for NDT operations to which assigned and for which certified. A Level III must also be generally familiar with appropriate NDT methods other than those for which specifically certified, as demonstrated by passing a Level III basic examination. Level III personnel are qualified to perform the following tasks:
- Develop, qualify, and approve procedures; establish and approve NDT methods and techniques to be used by Level I and Level II personnel.
- Interpret and evaluate test results in terms of applicable codes, standards, specifications, and procedures.
- Assist in establishing acceptance criteria where none are available, based on a practical background in applicable materials, fabrication, and product technology.
- In the methods for which certified, be responsible for, and capable of, training and examination of Level I and Level II personnel for certification in those methods.

Challenges
The major challenge facing NDT personnel is to learn all that can possibly be learned during the qualification processes. Another challenge involves developing the mindset that there is something else to learn each time the NDT method is used. There is no substitute for knowledge, and NDT personnel must be demanding of themselves. The work performed in the NDT field deserves the very best because of the direct effect of protecting life or endangering life.

Personnel Certification

It is important to understand the difference between two terms that are often confused within the field of NDT: qualification and certification. Qualification is a process that should take place before a person is certified. According to *SNT-TC-1A*, the qualification process for any NDT method should involve the following:

1. Training in the fundamental principles and applications of the method.
2. Experience in the application of the method under the guidance of a certified individual, referred to as on-the-job training.
3. Demonstrated ability to pass written and practical (hands-on) tests that prove a comprehensive understanding of the method and an ability to perform actual tests using the specific NDT method.
4. The ability to pass a vision test for visual acuity and color perception or shades of gray, as needed for the method.

The actual certification of a person in NDT to a Level I, Level II, or Level III is written testimony that the individual has been properly qualified. It should contain the name of the individual being certified, identification of the method and level of certification, the date, and the name of the person issuing the certification. Certification is meant to document the actual qualification of the individual.

Proper qualification and certification is extremely important because the process of testing performed by certified NDT personnel can have a direct impact on the health and safety of every person who will work on, in, or in proximity to the equipment or assemblies being tested. Poor work performed by unqualified personnel can cost lives.

Modern fabrication and manufacturing projects challenge the strength and endurance of the materials of construction. Preventive maintenance activities also present a challenge to NDT personnel.

The industries that depend on NDT cannot tolerate NDT personnel who are not adequately qualified, certified, and dedicated to good performance. Too much depends on the judgment of NDT personnel made in the work performed every day.

2

Electromagnetic Techniques and Magnetic Flux Leakage

Electromagnetic testing (ET) involves the application of electromagnetic energy to evaluate the condition of a range of test objects. As described in Chapter 1, *SNT-TC-1A* divides ET into three techniques: eddy current testing, alternating current field measurement, and remote field testing (RFT). In addition, the document regards magnetic flux leakage testing (MFL) as a separate nondestructive test method. Since principles of flux leakage are common to both ET and MFL, it is included in this *Classroom Training Book.*

- Eddy current testing: uses an alternating current (AC) magnetic field to induce circular currents that are affected by fissures, constrictions, or other discontinuities in a test object. It is a useful technique for detecting the wall thickness of thin materials as well as detecting localized discontinuities.
- Alternating current field measurement: involves the induction of a locally uniform current into a test object and measuring the magnetic flux density above the test object surface. It was originally designed for the detection and sizing of fatigue cracks.
- Remote field testing: can inspect regions not only near the probe but also throughout the thickness of the material. A major application is the testing of thick-walled ferromagnetic tubing where remote field testing shows equal sensitivity to internal and external discontinuities.
- Magnetic flux leakage testing: provides an assessment of the integrity of ferromagnetic material. The magnetic flux leakage can be detected using noncontact sensors. Cracks in the formative stage are highly detectable and normally exhibit well-defined indications.

This chapter offers an introduction to the three ET techniques and the MFL method, each of which will be discussed in more detail in subsequent chapters. Whereas eddy current testing is typically used on nonferromagnetic parts, if used properly, either of the other two ET techniques or MFL can provide useful information about ferromagnetic components. The deciding factor of one over the other is the type of material, part size or geometry, and the type and size of discontinuities the customer needs to be able to detect. The distinctive nature of each technique or method stems from differences in excitation frequencies, the nature of the transducers used, and the signal analysis techniques for characterizing the state of the test object. We start with eddy current testing, since this was the first of the three ET techniques to be developed.

Eddy Current Testing

Eddy current testing is most commonly used for detection of surface or near-surface discontinuities in nonferromagnetic materials. In materials with little or no permeability, eddy current testing is effective to approximately 0.2 in. (5 mm) below the test surface. For material thicknesses of greater than 0.2 in. (5 mm), special probes and/or electronics packages are needed to improve the performance of the eddy current technique.

Although there are some highly successful applications for eddy current tests on ferritic materials, eddy current has limited ability to provide subsurface discontinuity detection in ferromagnetic alloys. Surface-crack detection in ferromagnetic materials, especially for weld inspection, is a viable eddy current process when the right technology is applied. Eddy current is often more sensitive than either magnetic particle testing (MT) or liquid penetrant testing (PT) in this role. In many cases, eddy current does a better job with improved sensitivity in much less time.

As discussed in Chapter 1, eddy current testing is based on the fact that, when a coil excited by an alternating current is brought close to a material, the terminal impedance of the coil changes. The change is associated with the fact that the primary field set up by the eddy current coil induces eddy currents within the electrically conductive specimen. The direction of the induced eddy currents and, consequently, the secondary field generated by these currents, oppose the change in the primary field, as shown in Figure 1.

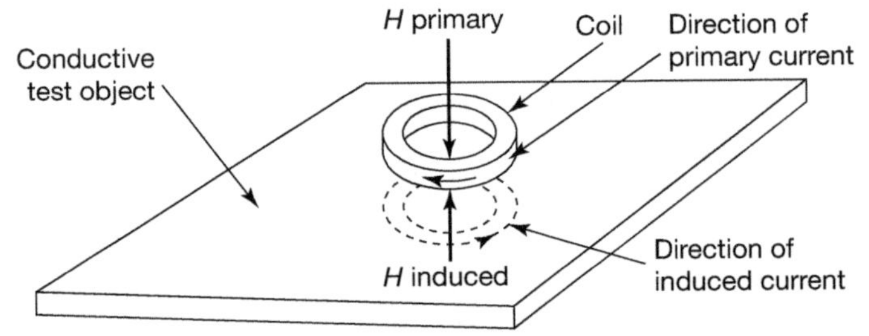

Figure 1. Alternating current coil over a conductive test object, showing opposite direction of primary and induced currents.

The presence of a discontinuity or heterogeneous material in the test object causes a redistribution of the normal eddy current flow. When the eddy currents are altered by the material properties, the electrical properties of the coil are affected. Those changes are detected and displayed on an output device.

The variations in coil impedance caused by discontinuities in the test object are often very small in comparison with the value of the coil impedance at rest. Detection and measurement of these small changes is often accomplished using bridge circuits. In this regard, there is a useful distinction between two kinds of probe coils: absolute coils and differential coils. An absolute coil responds to the electromagnetic properties of the test object in the magnetic field of the coil without comparison to the response of a second coil. Differential coils are two or more coils connected in such a way that electromagnetic differences in the regions beneath the coils will cause an imbalance between them to be signaled. An example of a differential eddy current probe is shown in Figure 2.

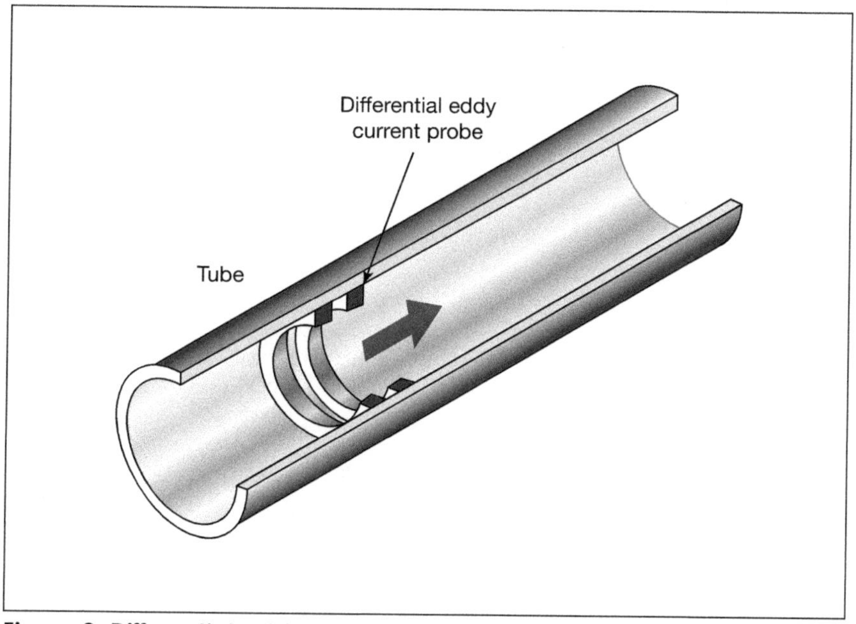

Figure 2. Differential eddy current probe for inspecting tubes from the inside.

Reference Standards

An important requirement for successful ET is the use of an accurate reference standard for equipment standardization. The reference standard is used to adjust the electromagnetic equipment's sensitivity to various test object parameters (cracks, surface roughness, conductivity/ permeability variations, and other material conditions). Success in electromagnetic testing depends on the proper use of available reference standards. The development and use of eddy current reference standards requires a thorough understanding of the test to be performed. An example is shown in Figure 3.

Figure 3a shows a custom-made 0.003 in. (0.076 mm) deep saw cut at 31× magnification in a 0.941 in. (2.39 cm) diameter steel bar that serves as a reference standard for an electromagnetic test using a rotary system to spin the bar. A microscopic full-length scratch was found in the sample supplied by the customer, and the notch was placed on the opposite side of the bar. The screen display in Figure 3b shows data from the saw cut and the signal from the preexisting scratch on the test sample.

(a)

(b)

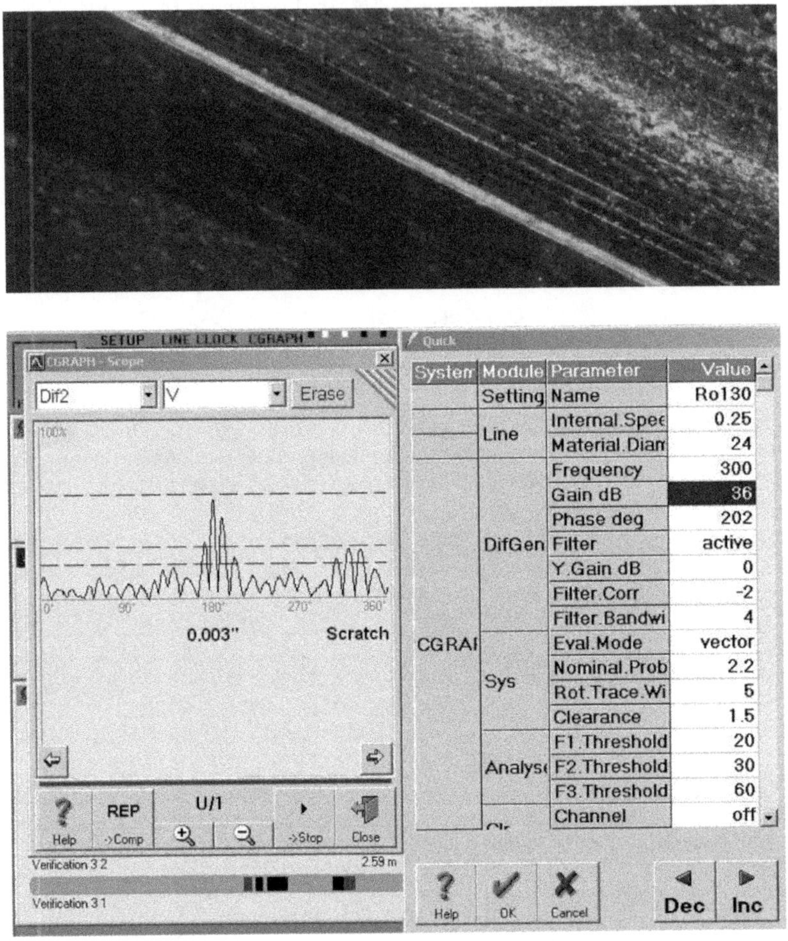

Figure 3. Eddy current reference standard: (a) saw cut in a steel bar; (b) screen display showing data.

Digital Data Displays

Digital data are generally displayed in a complex plane presentation with supporting strip chart and C-scan displays, as required by the application. The point described by the in-phase and out-of-phase components of the signal is typically displayed as a "flying dot," and the digital capabilities of the instrument allow variable persistence, centering of the dot, rotation of the signal, and scaling of the display. Digital systems allow setup of standardization curves constructed from stored data and automated analysis of signals as compared to these curves.

Alternating Current Field Measurement

The primary application of alternating current field measurement is the detection and sizing of surface-breaking cracks in ferromagnetic welds using an AC power source. Compared to magnetic particle testing and liquid penetrant testing, alternating current field measurement offers the following advantages:

- Works through nonconductive coatings up to 0.4 in. (10 mm) thick so there is no need to remove and then reapply paint or to clean off rust.
- Provides information on depth as well as length, saving time on removing discontinuities of insignificant depth.
- Relatively insensitive to material property changes, so ideal for inspecting all welds.
- Relatively insensitive to probe liftoff, allowing deployment through coatings and on rough surfaces.
- Allows depth sizing of discontinuities up to about 1 in. (25 mm) depending on the probe type.

Alternating current field measurement has been developed from the alternating current potential drop technique. Alternating current potential drop uses current injection and contact potential drop probes. This technique requires extensive surface preparation of the weld under examination and can be used to produce crack depth measurements.

The contact probes previously used in alternating current potential drop have been changed to noncontact magnetic field sensitive coils, and the current is induced into the test object. The models developed in alternating current potential drop for mapping surface magnetic fields and electric currents have been utilized in alternating current field measurement.

The alternating current field measurement technique in its simplest form uses a handheld probe containing a uniform field induction system and two magnetic field sensors.

The induced AC is generated in a limited region of the test object where the alternating electric current is considered to be linear. In this region, a magnetic field is produced that is also linear. Any disturbances in this region produced by surface discontinuities will affect the components of this linear magnetic field. Two or more air-wound coils mounted with orthogonal axes within a probe will detect these disturbances. This is the foundation of alternating current field measurement, which differentiates it from eddy current testing.

Alternating Current Field Measurement Principles

Figure 4 shows the basic principles of the technique. With no discontinuity present and a uniform current flowing in the Y direction, the magnetic field is uniform in the X direction perpendicular to the current flow, while the other components are zero. The presence of a discontinuity diverts current away from the deepest parts and concentrates it near the ends of a crack. The effect of this is to produce strong peaks and troughs in B_z above the ends of the crack, while B_x shows a broad dip along the whole discontinuity with amplitude related to the depth.

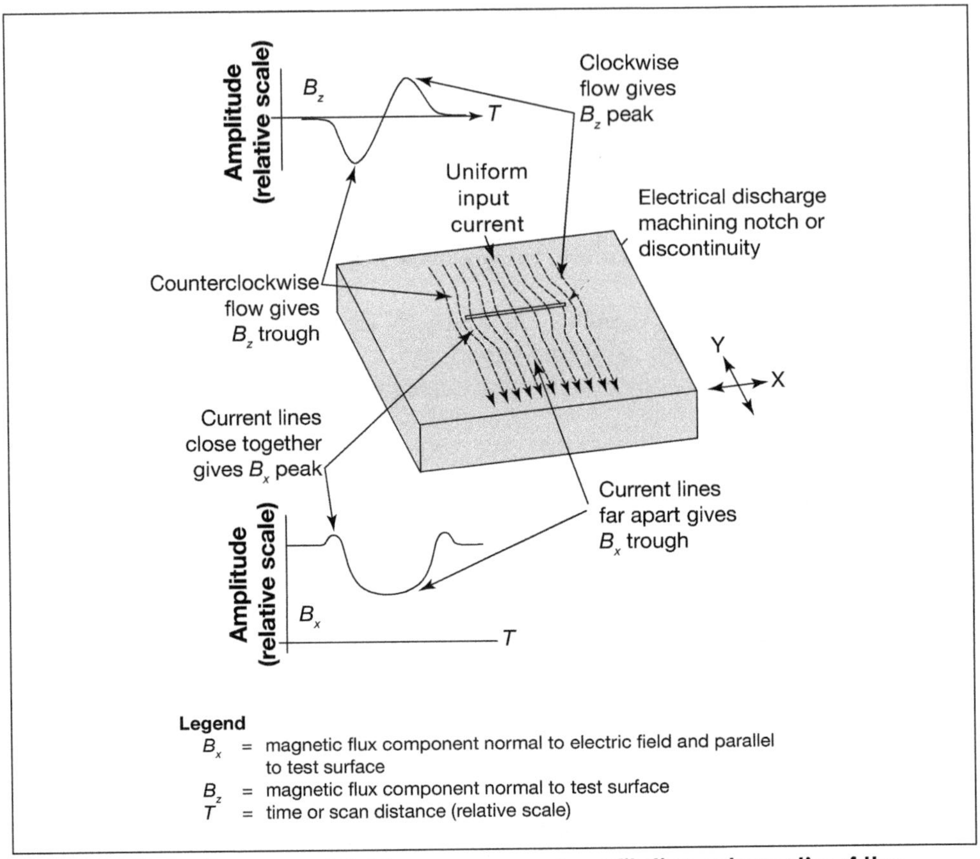

Figure 4. Alternating current field measurement qualitative schematic of the magnetic forces above a notch.

Alternating Current Field Measurement Applications

Inspection companies and owners of fabricated components use the alternating current field measurement technique for weld inspection in petrochemical process plants, pharmaceutical plants, offshore well structures, highway bridges, and roller coasters. Originally introduced to the offshore industry for subsea weld inspection, the use of alternating current field measurement has now broadened to include inspection of pressure vessels, process piping, and drill-pipe threads and risers. Recent developments have included automated and semi-automated systems to reduce the reliance on operators and the use of array technology to increase inspection speeds.

Alternating current field measurement can be used for the inspection of nonferromagnetic materials but is less effective in this role. The relative discontinuity sensitivity in nonferritic materials with alternating current field measurement is dramatically reduced. This is in sharp contrast to standard eddy current theory. It should also be noted that volumetric discontinuities, such as corrosion pitting or porosity, give much weaker signals than planar discontinuities, so it is not recommended that alternating current field measurement be used in this role.

Remote Field Testing

Remote field testing (RFT) is a versatile technique that can be used effectively to test steel tubes and pipes. RFT is popular because of its ability to inspect regions not only near the probe but also throughout the thickness of the material. This characteristic is especially valuable for the testing of thick-walled ferromagnetic tubing, where the technique shows high sensitivity to inside and outside surface pipe wall discontinuities. RFT often does not require cleaning of the pipe and is not sensitive to internal coatings. Since 1970, RFT has grown into a mature and recognized nondestructive testing technology.

There are reference materials that talk about "remote field eddy current," but RFT should not be looked at as a typical eddy current test. To prevent confusion on the range of applications and material test situations, the attempt is being made to phase out that particular terminology. Both ASTM and ASME have RFT listed as a specific technique within ET. Thus, *SNT-TC-1A* also includes it as an electromagnetic technique. For the purpose of generic discussion, we will be looking at RFT as it applies to the inspection of ferromagnetic tubing in heat exchangers.

Remote Field Principles

RFT is an electromagnetic technique that utilizes an AC excitation source. This AC electromagnetic energy travels along the tube wall for some distance in both directions from an exciter coil. The distribution of the primary field is dependent upon the magnetic properties of the tube, the tube wall thickness, and the presence of surrounding support structures. The transmitted field may be affected by anomalies within the tube wall or when it encounters support structures on the tube outside diameter (OD). The changes in the strength (amplitude) and phase shift (or phase angle) of the received signal are measured a few tube diameters away from the exciter coil.

Special hybrid (driver/pickup) coils are necessary to perform RFT inspections. Because of the need for significant spacing between the exciter coil(s) and the receiver or pickup coil(s), the probes tend to be longer than the typical eddy current probe. (See Figure 5.)

The high magnetic permeability of ferromagnetic materials dramatically impacts standard eddy current testing techniques. Some eddy current techniques attempt to compensate for and/or suppress the permeability effects by the use of strong magnets or DC-driven saturation coils. The RFT process requires no magnetic saturation. Instead, it makes use of the natural tendency of ferromagnetic materials to channel magnetic energy. Like the keeper of a horseshoe magnet, the magnetic lines of flux from the exciter coil take the path of least reluctance. They flow down the tube wall, which acts as a waveguide, for a considerable distance.

At distances in excess of two tube diameters from the internal exciter coil, the flux field becomes homogenous and the passive receiver coils, positioned two to three tube diameters away from the exciter, receive practically all of their energy from the flux in the tube wall. The direct field from the exciter is almost completely attenuated (or absorbed)

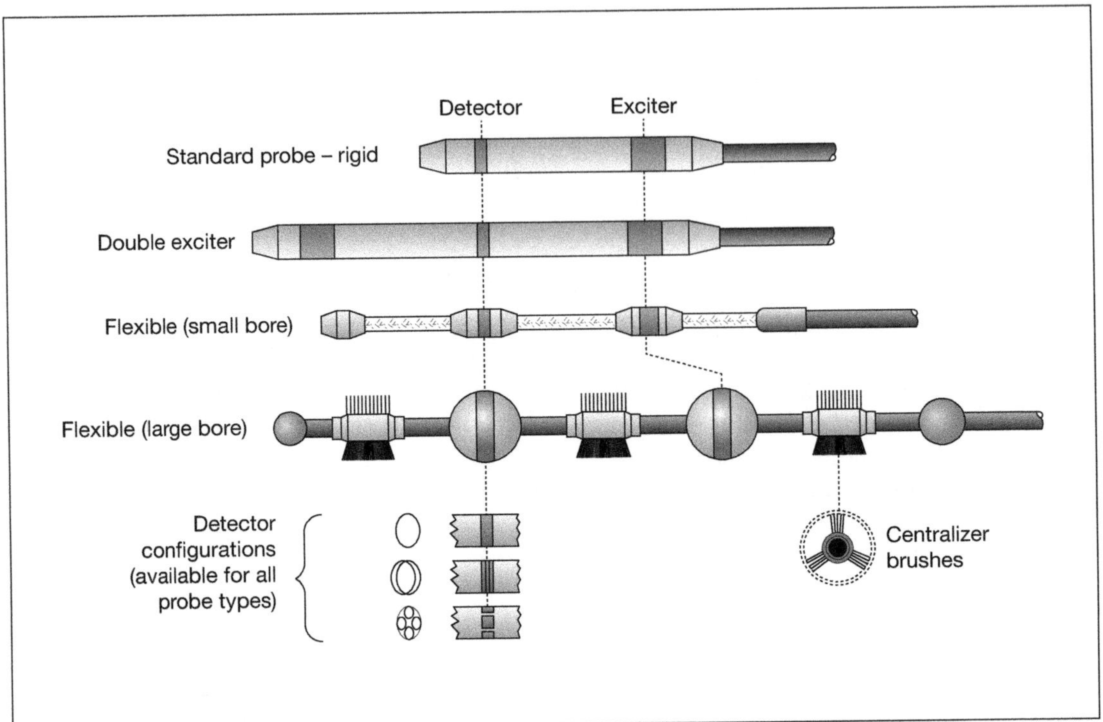

Figure 5. Remote field testing probe types (top to bottom): larger diameter tubing (with either single or dual exciters), smaller diameter tubing, and boiler tubing. (Courtesy of Russell Tech.)

by the tube wall, and the external field is actually stronger than the remaining magnetic field inside the tube.

Remote Field Energy Zones

For a remote field probe, there are two distinct sensing zones with a transition zone between them (Figure 6). In order, the zones are the direct field zone, the transition zone, and the remote field zone. As the detector coil distance from the exciter coil is increased, the dominant field energy changes from direct coupled energy (between the exciter and detector coils, inside the tube) to energy that is coupled to the detector coil primarily by transmission through the tube wall. Between these two distinct zones, there is a transition zone where the direct coupled energy and the indirect coupled energy are comparable in magnitude. The location of the transition zone changes with frequency, wall thickness, permeability, and conductivity.

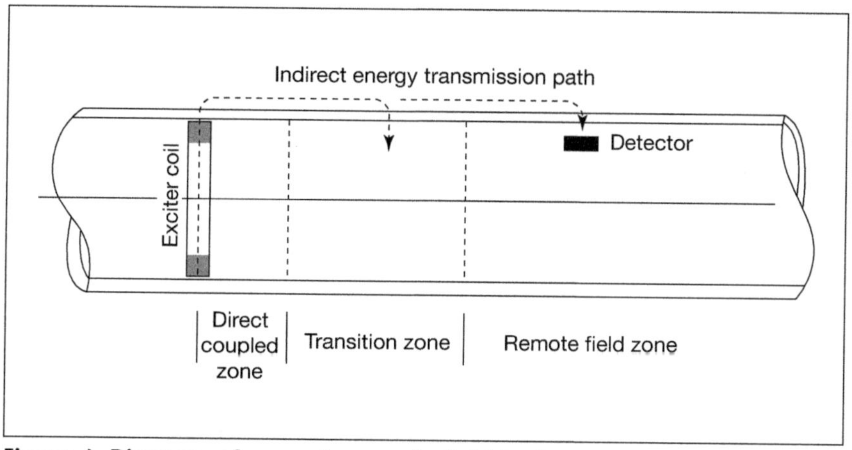

Figure 6. Diagram of zones in remote field testing.

Magnetic Flux Leakage

The primary application of magnetic flux leakage testing (MFL) is ferromagnetic materials, including tubing, plate, wire, oilfield tubulars, and pipelines. In the petrochemical world, MFL is used predominantly for tank floor inspection (Figure 7). The power source is provided by permanent magnets or DC or AC coils.

MFL has been extensively used in the pipe inspection industry. This entails the introduction of a moving DC magnetic field into a ferromagnetic test piece. Any localized, normally surface-breaking discontinuities that lie within the inspection zone will cause the field to bend or leak and extend above the surface at that point. These flux lines cut across a moving coil or sensor and are used to detect the leakage field.

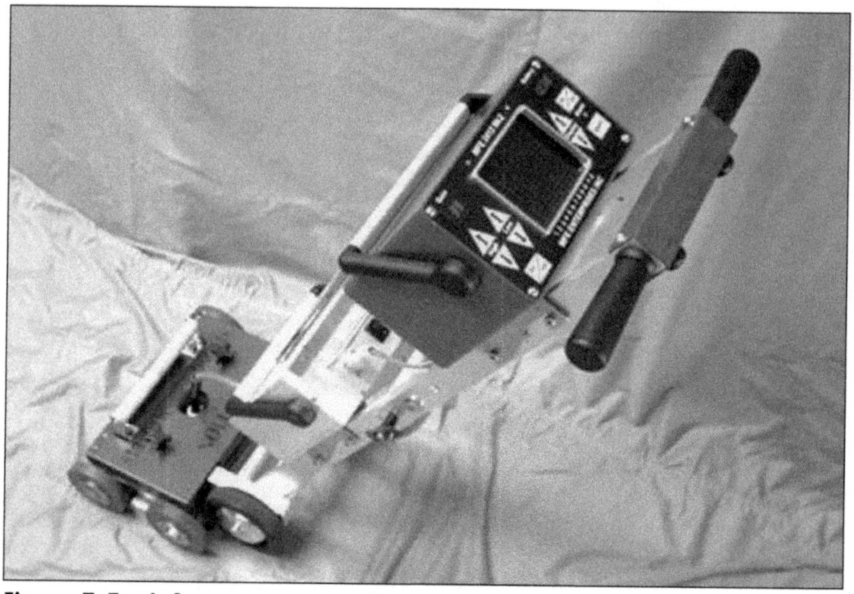

Figure 7. Tank floor scanner system.

MFL Inspection

In many piping and tubing tests, MFL is used to look for general corrosion. The locally thinned area puts a higher magnetic flux distribution in the space nearer to the flux detection device. This relative increase in field strength can be measured. MFL performs best with the detection of volumetric discontinuities rather than small pits or cracks. Any discontinuity with its major axis parallel to the direction of the flux flowing in the material has little chance of being detected using this method.

The MFL probe pull speed must be maintained at a fairly constant rate. If not, then the accuracy of the test is decreased even further.

Pipeline inspections are performed with what are called smart pigs. These devices can carry out multiple nondestructive tests simultaneously. However, the most common test performed is MFL.

Large inspection pigs use MFL in order to detect internal or external corrosion. They use a circumferential array of detectors positioned between the poles of strong permanent magnets to magnetize the pipe wall to near saturation flux density. Abnormalities in the pipe wall, such as corrosion pits, create magnetic flux leakage near the pipe's surface. The flux leakage may be detected by hall effect probes or passive induction coils. (See Figure 8.)

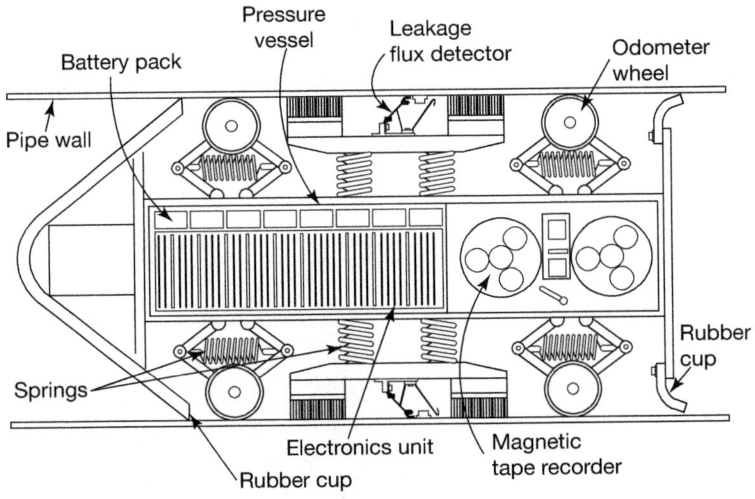

Figure 8. Sketch of magnetic flux leakage testing pipe inspection device. (Courtesy of Queen's University, AMG)

The demands now being placed on magnetic inspection tools are shifting from the mere detection, location, and classification of pipeline discontinuities to the accurate measurements of discontinuity size and geometry. Modern, high-resolution MFL tools are capable of giving very detailed signals. However, converting these signals to accurate estimates of size requires considerable expertise, as well as a detailed understanding of the effects of inspection conditions and the magnetic behavior of the type of pipeline steel used.

Alternating Current versus Direct Current

AC magnetization is more suitable for detection of outer surface discontinuities because it concentrates the magnetic flux at the surface. For equal magnetizing forces, an AC field is better for detecting outside surface imperfections, but a DC field is better for detecting imperfections below the surface.

When tests for both outer surface and inner surface discontinuities are necessary, it may be best to test first for outer surface discontinuities with an AC field, then for inner surface discontinuities with a DC field.

3

Electromagnetic Theory

Electric Current

Every atom consists of three types of fundamental particles: electrons, protons, and neutrons. Protons and neutrons reside in the nucleus at the center of the atom, while electrons, according to the model developed by Niels Bohr, revolve around the nucleus in fixed orbits at different energy levels (Figure 1). (With the development of quantum theory, this model has since been modified to envision electrons as revolving about the nucleus of an atom in a region forming an electron cloud.) Electrons possess a negative charge, whereas protons are positively charged and neutrons have no charge. The unit of charge is the coulomb (C). One electron or one proton has a 1.6×10^{-19} C charge, negative or positive, respectively, and denoted by e.

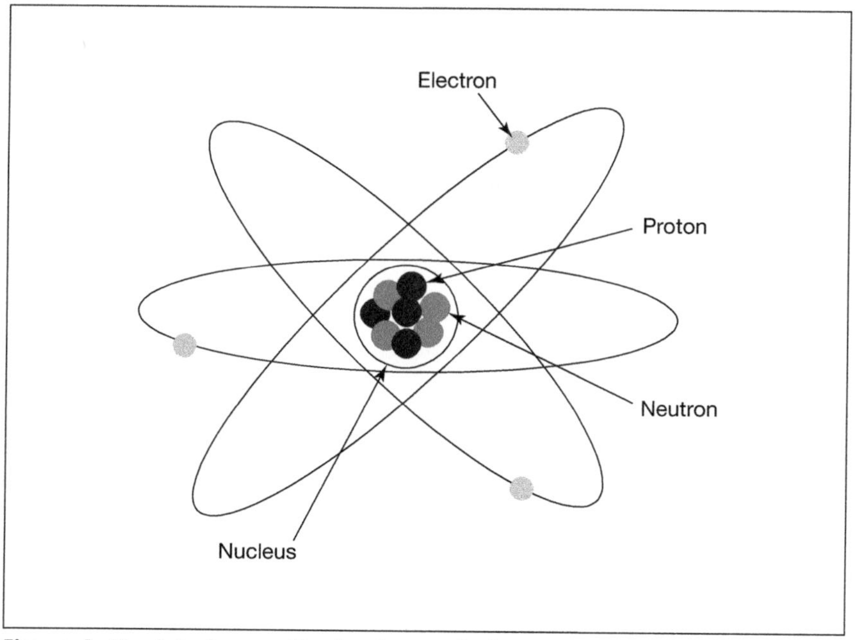

Figure 1. Model of a positively charged atom.

A conductor is any material that is capable of carrying electrical current. Some materials are conductors, others are not. Whether a material can conduct electricity or not depends on whether there are electrons that are freed from their parent atoms so that they can move relatively freely through the material's crystal structure when acted upon by an externally applied voltage.

The quantity of charge (q) on an object is equal to the number of elementary charges on the object (n) multiplied by the elementary charge (e).

(Eq. 1) $$q = ne$$

Electric current (I) is defined as the movement of electrons through a conductor. The unit of current is the ampere. Its quantitative value is established as the flow of 6.25×10^{18} electrons per second past a given point in the circuit. Electric current may also be expressed as the rate of flow of a quantity of electric charge (q) through a conducting medium with respect to time (t).

(Eq. 2) $$I = \frac{q}{t}$$

Its unit is coulomb per second (C/s) or ampere (A).

(Eq. 3) $$\text{A} = \text{C} / \text{s}$$

Conversely, the coulomb is the quantity of electricity carried in 1 second (s) by a current of 1 A.

Problem 1

A total of 100 C of charge is transferred through a conductor in 50 s. What is the current?

Given

$q = 100$ C
$t = 50$ s

Solution

$$\begin{aligned} I &= \frac{q}{t} \\ &= \frac{100\ \text{C}}{50\ \text{s}} \\ &= 2\ \text{A} \end{aligned}$$

Electromagnetism

Electron movement is responsible for current flow. The Danish physicist Hans Christian Örsted explained that when electrons flow through a conductor, a magnetic field gets produced around that conductor. The direction of the produced magnetic field can be determined through the right-hand rule (Figure 2), in which the thumb points in the direction of current, and the curled fingers indicate the direction of lines of magnetic field. The lines of flux of the magnetic field can be imagined as spreading out in two directions on a sheet of paper intersecting the flow of magnetism (Figure 3).

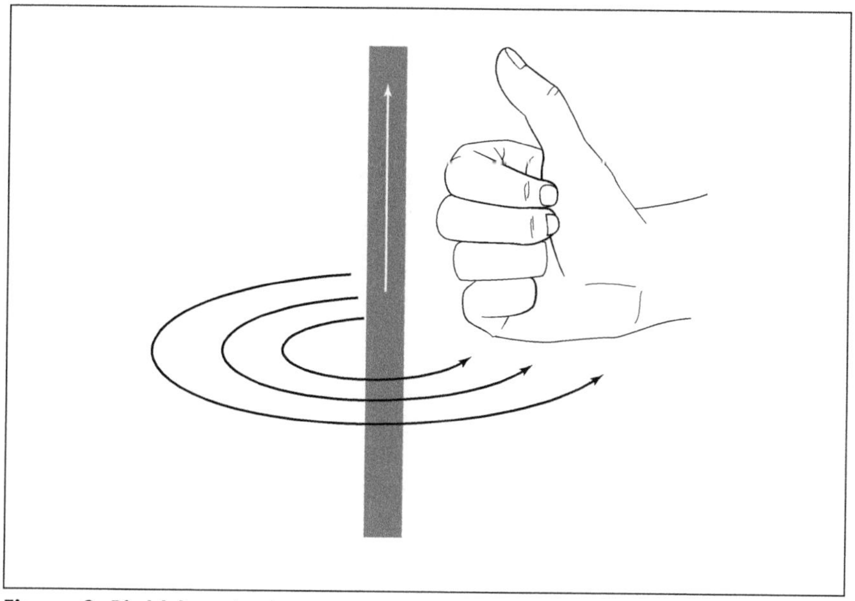

Figure 2. Right-hand rule.

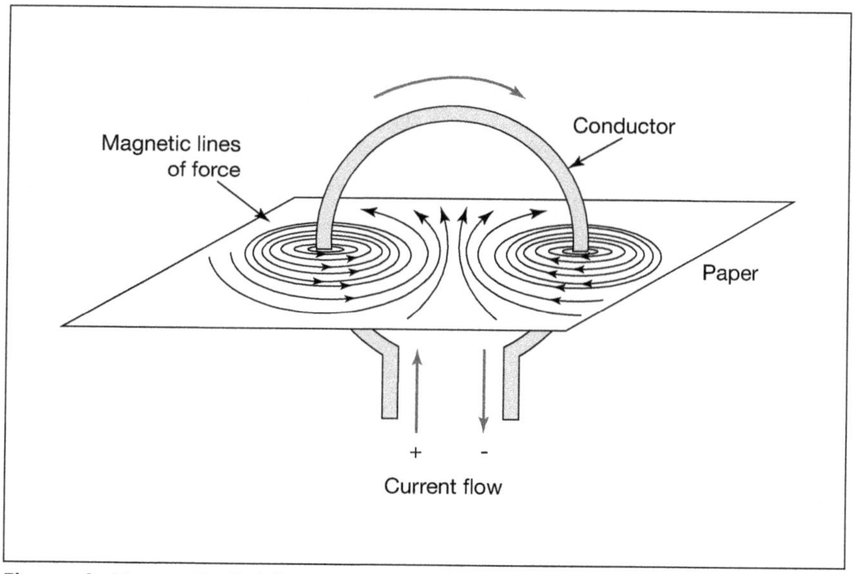

Figure 3. Magnetic field.

The magnetic field generated through the flow of current is called the primary field. The field produced around this loop of wire, referred to as a coil, is directly proportional to the magnitude of the applied current, rate of change of the current or frequency, and the coil parameters. Coil parameters include the size and shape of the coil, the amount of wire used to fill the coil geometry, the electrical properties of the wire itself, and any shielding or core materials used to house the coil.

Magnetic Flux

The number of magnetic lines of forces passing per unit area perpendicular to the direction of a magnetic field is the magnetic flux (ϕ). Magnetic flux is the product of the average magnetic field (B) and the area (A) through which the magnetic field penetrates perpendicularly.

(Eq. 4) $$\phi = BA$$

Michael Faraday explained that if the primary magnetic field varies, it will generate a secondary current in any nearby conductive material. This secondary current, generated through the varying magnetic fields, is referred to as an eddy current. Note: Flux leakage theory is discussed in more detail in Chapter 4.

Generation of Eddy Currents

Each ET technique uses some form of electromagnetic induction. That means that eddy currents are active in either the material or the sensors used to monitor them. An eddy current test system includes a test coil through which varying or alternating current (AC) is passed. A varying current flowing in a test coil produces a varying electromagnetic field around the coil known as the primary field, as noted above.

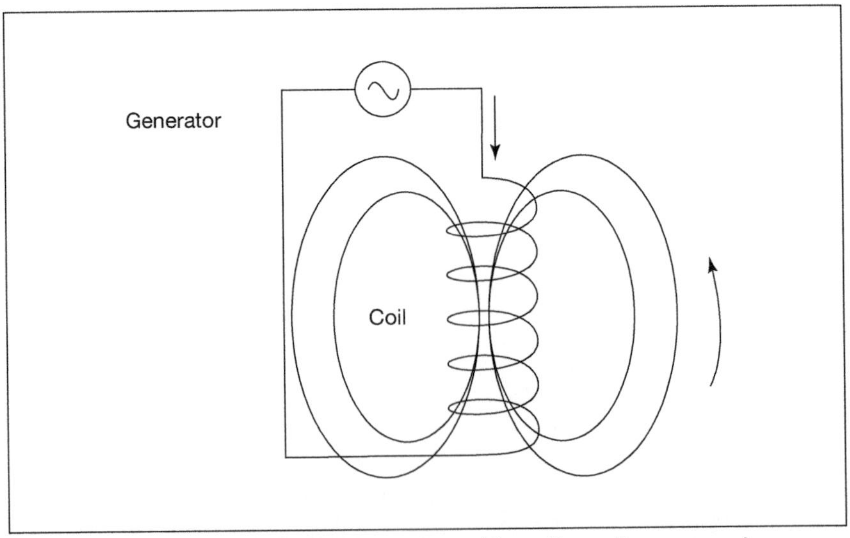

Figure 4. Electromagnetic field produced by alternating current.

Figure 4 presents a schematic view of an excited test coil. The electromagnetic field produced around the unloaded test coil can be described as decreasing in intensity with distance from the coil and also varying across the coil's cross section. The electromagnetic field is most intense near the coil's surface.

Electromotive Force

Electromotive force (emf) is the electrical energy derived from mechanical, chemical, or other forms of energy that must be applied across the material to force the electrons to move. The unit of measurement of emf is the volt (V).

Michael Faraday's major contribution to the understanding of emf was the discovery of electromagnetic induction. His work can be summarized in two laws.

Faraday's First Law

Whenever a conductor is placed in a varying magnetic field, an emf gets induced across the conductor, and if the conductor is a closed circuit, induced current flows through it. The emf setup in the coil is called induced emf, and the current thus produced is termed an induced current.

This law is demonstrated in Figure 5. Coil A is connected to a battery through a switch. A second coil (B) connected to a galvanometer is nearby. (A galvanometer is an instrument for detecting and measuring small electric currents.) When the switch is closed producing a current in coil A in the direction shown, a momentary current is induced in coil B in a direction (→ a) opposite to that in A. If the switch is now opened, a momentary current will appear in coil B having the direction of (→ b).

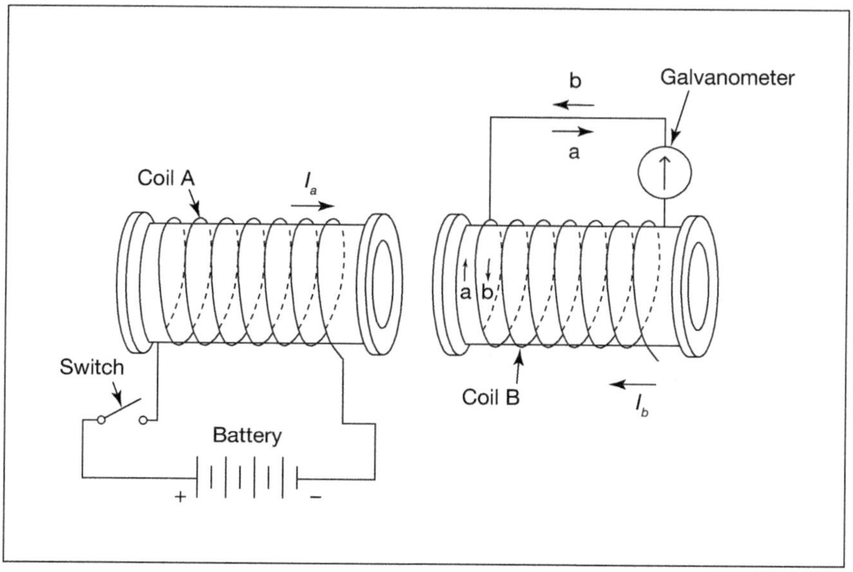

Figure 5. Induced current.

In each case, current flows in coil B only while the current in coil A is changing. (I = current flowing in the coil.)

Some materials, due to their molecular structure, require more energy than others to cause electrons to move. These are said to have more resistance to the flow of electrical current. The amount of resistance in a material is the factor that limits the amount of current that flows through the material for a given applied emf.

Faraday's Second Law

The induced emf (E) is proportional to the rate of change of the magnetic flux ($\Delta\phi/\Delta t$). Thus, the emf (voltage) induced in coil B of Figure 5 can be expressed as:

(Eq. 5)
$$E = -N\frac{\Delta\phi}{\Delta t}$$

where

E is the instantaneous induced voltage

N is the number of turns in the coil

$\Delta\phi$ / Δt is the time rate of change of magnetic flux through the coil in webers per second (Wb/s)

Lenz's law, named after Russian physicist Heinrich Lenz, explains the existence of the negative (minus) sign. Lenz's law states that induced current produced by a varying magnetic field will flow in such a way that the induced current will create its own magnetic field that opposes the magnetic field that created it.

In the International System of Units (SI), the weber (Wb) replaces the traditional unit of the maxwell (Mx) as a measure of magnetic flux, such that:

(Eq. 6)
$$\begin{aligned} 1\text{ Mx} &= 10^{-8}\text{ Wb} \\ &= 0.01\ \mu\text{Wb} \\ &= 10\text{ nWb} \end{aligned}$$

Resistance

In an AC circuit containing only resistance, the resistance simply limits the amount of current that flows through the circuit. It does not change the phase relationship between the voltage and the current. The current is exactly in phase with the voltage. Resistance is present in all circuits. The total resistance in a circuit includes the resistance of the wiring as well as the resistance of the coil.

With all other factors held constant, changes in resistance will affect the impedance of the circuit. As the resistance increases, the impedance increases. Note: The concept of phase is discussed in more detail in later sections, titled "Self Inductance" and "Impedance".

Ohm's Law

The unit of resistance is the ohm (Ω). Its value is defined as the resistance through which an electromotive force (emf) of 1 V will produce a current of 1 A. In a direct current (DC) circuit, the voltage, current, and resistance are related to each other through the mathematical expression known as Ohm's law. Ohm's law explains that the potential difference (E) across the conductor is proportional to the current (I) flowing through it.

(Eq. 7a) $$E = I \times R$$

or

(Eq. 7b) $$I = \frac{E}{R}$$

where

E is voltage (V)
I is current (A)
R is resistance (Ω)

Problem 2

What is the value of the current flowing in a 5 Ω resistor connected across a 10 V battery?

Given

E = 10 V
R = 5 Ω

Solution

$$I = \frac{10\ \text{V}}{5\ \Omega}$$
$$I = 2\ \text{A}$$

The current flowing across a 5 Ω resistor across a 10 V battery is 2 A.

Specific Resistance

The resistance of a coil is determined by the length of wire used to wind the coil and the unique characteristics of that wire:

(Eq. 8) $$R = \rho\frac{l}{A}$$

where

R is resistance (Ω)
ρ is specific resistance in circular mil ohm/ft (Ω·m)
A is the area, in circular mil (m^2)
l is the length, in feet (m)

In addition, specific resistance of a material may change with temperature, with the standard temperature usually specified at 20 °C.

Problem 3

The resistance of a 10 ft (3.048 m) length of 40 gauge wire (area of 40 gauge wire = 9.888 circular mil [0.000000005 m^2]) with a specific resistance of 10.4 circular mil ohm/ft (0.000000017 Ω·m) at 20 °C would be found as follows.

Given

ρ = 10.4 circular mil ohm/ft (0.000000017 Ω·m)
A = 9.888 circular mil (0.000000005 m^2)
l = 10 ft (3.048 m)

Solution

In traditional (imperial) units:

$$R = \frac{10.4 \text{ circular mil } \Omega / \text{ft} \times 10 \text{ ft}}{9.888 \text{ circular mil}}$$
$$= \frac{104 \text{ circular mil } \Omega}{9.888 \text{ circular mil}}$$
$$= 10.518\ \Omega$$

In SI units:

$$R = \frac{0.000000017\ \Omega \bullet \text{m} \times 3.048 \text{ m}}{0.00000005 \text{ m}^2}$$
$$= \frac{0.0000000518\ \Omega \bullet \text{m}^2}{0.000000005 \text{ m}^2}$$
$$= 10.36\ \Omega$$

The difference in results is explained by the inexact rounding of conversion factors to large decimal places in the two sets of givens.

Conductance

The opposite of resistance is conductance. The lower the conductance, the higher the resistance. The symbol of conductance is *G*. In SI, conductance is expressed in siemens (S); informally, it is expressed in mhos. Siemens and mhos are equivalent units.

(Eq. 9) $$1 \text{ S} = 1 \text{ mho}$$

The "mho" is "ohm" spelled backward. This is because G is the reciprocal of R (expressed as ohms or Ω), as follows.

(Eq. 10) $$G = \frac{1}{R}$$

As we continue to study the electromagnetic method we will keep coming back to how the material being tested reacts to the energies being used. The material properties that must be understood in every electromagnetic test are:

1. Conductivity, σ (sigma): an electrical property; how well this material allows current to flow through it.
2. Resistivity, ρ (rho): an electrical property; how much opposition the material provides to the current flowing through it.
3. Permeability, μ (mu): a magnetic property; how the material responds to the magnetic lines of force that are trying to move through it.

If we can understand the effect these material properties will have on the test, then our ability to properly interpret the test results will be much greater.

Alternating Current

When a coil of wire is placed in the open end of a magnet between the north and south poles and given a spin, electricity is induced in the coil. The current produced does not travel in the same direction through the coil at all times, nor is it of a constant value. Instead, the current starts out at zero, rises to a maximum value, decreases to zero, rises to a maximum value in the opposite direction, and then returns to zero. This cycle repeats itself as long as the coil keeps spinning. Note that one revolution of the coil produces one cycle of current (Figure 6).

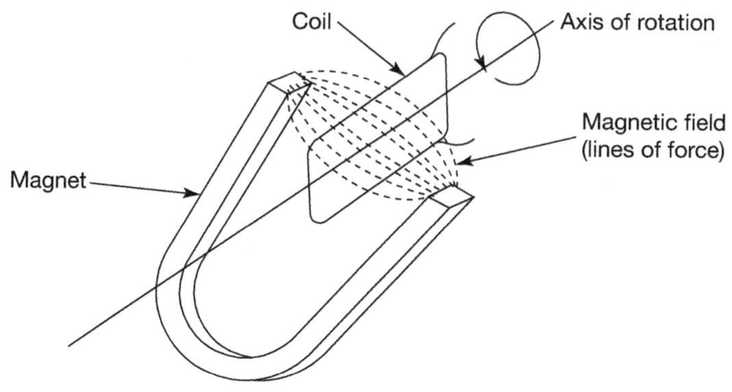

Figure 6. Induction by movement of a coil through a magnetic field.

Sine Wave

A sine wave is the form commonly produced by AC generators. Since one turn (360° rotation) of the generator coil produces one cycle of the sine wave, the sine wave can be marked into corresponding degrees of rotation, as shown in Figure 7. This method of designating positions on the sine wave serves as an effective way of showing the timing between specific occurrences.

Frequency

The frequency of AC is defined as the number of cycles of current that occur in one second. One complete cycle is shown in Figure 7. Its unit is the hertz, 1 Hz being one cycle per second. Current at 60 cycles per second has a frequency of 60 Hz; at 50 cycles per second, the frequency is 50 Hz.

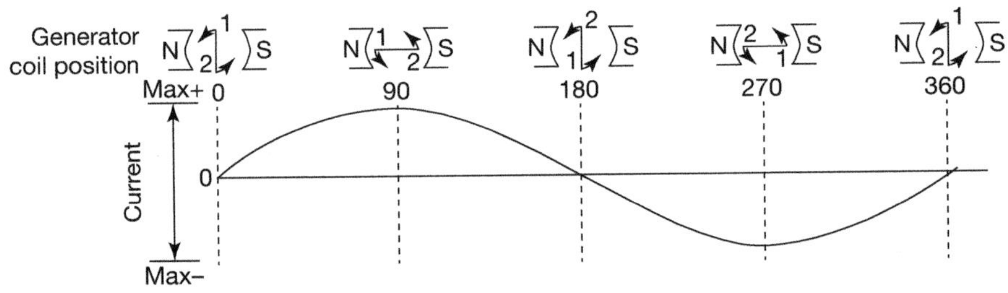

Figure 7. Generator coil position versus current produced.

Self Inductance

The process of generating an induced magnetic field and induced current is called inductance. When working with the induction of current in a secondary coil, American scientist Joseph Henry went even further than Faraday in that he discovered that the changing magnetic field also induced a current in the primary coil that opposed the original current. This opposing current was the result of the magnetic field cutting across windings in the primary coil. The magnetic field created by each turn of wire in the coil affected all of the other turns in the same coil. Thus, in a coil, the varying magnetic field generates an induced emf that opposes the change through which it was generated. This process of self generation of induced emf is known as the principle of self induction.

Figure 8 shows an AC source connected to a coil. A voltmeter is provided to measure the voltage applied to the coil, and an ammeter is provided to measure the current through the coil. If the instantaneous values of voltage and current are plotted on a graph, the current is found to lag behind the voltage in time as shown in the lower portion of the figure. The angle between the two curves is called the phase angle or phase lag.

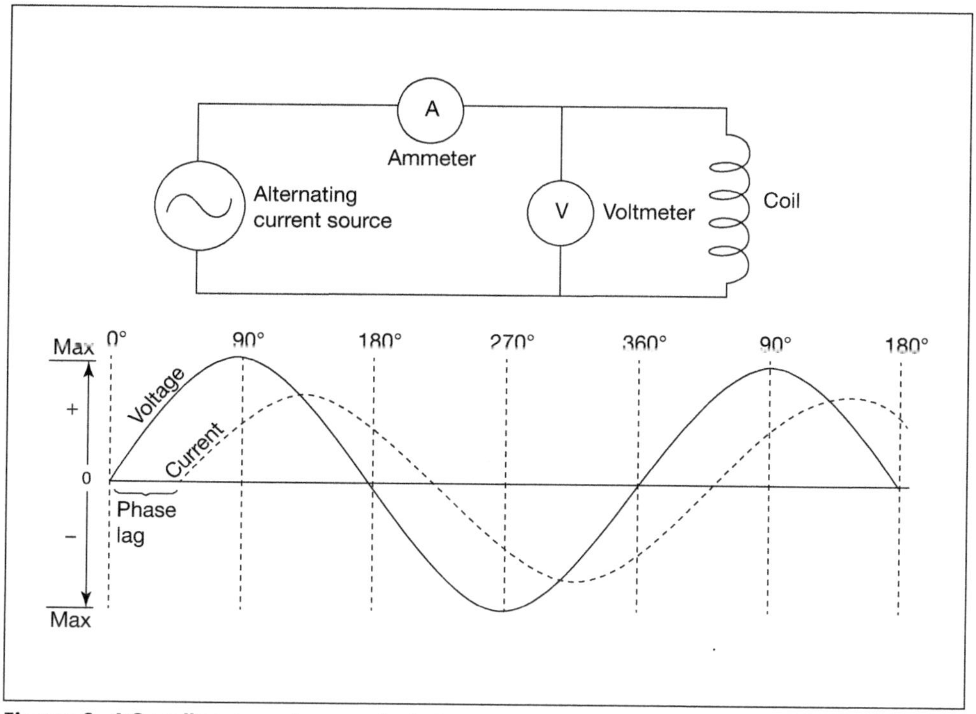

Figure 8. AC voltage and current plot.

Inductive Reactance

The opposition to changes in AC flow through a coil is called inductive reactance and is designated by the letters X_L. The inductive reactance of any coil is also a function of the frequency of the alternating current. Since the higher frequencies cause the magnetic field to change more rapidly, the inductive reactance increases as the frequency increases (assuming a constant alternating current voltage). The increase in the inductive reactance due to the increase in frequency causes the current through the coil to be reduced, thereby reducing the strength of the magnetic field of the coil. This is explained in Equation 11:

(Eq. 11a) $$X_L = 2\pi fL$$

or

(Eq. 11b) $$X_L = \omega L$$

where

X_L is inductive reactance (Ω)
f is frequency (Hz)
L is inductance (H)
ω is $2\pi f$ = angular frequency (rad/s)

It has been determined that in an AC circuit containing only inductive reactance (no resistance), the current will lag behind the voltage by exactly 90°.

Note: The henry (H) is the SI unit of inductance. The inductance of an electric circuit is 1 H when an electric current that is alternating at 1 A/s results in an emf of 1 V across the inductor.

Problem 4

What is the inductive reactance of a coil with an inductance of 5 microhenries (μH) at 60 kHz?

Given

$L = 5$ μH

$f = 60$ kHz

Solution

$$\begin{aligned} X_L &= [(2 \times 3.14) \times 60\ 000 \text{ Hz}] \times 0.000005 \text{ H} \\ &= (6.28 \times 60\ 000) \times 0.000005 \\ &= 376\ 800 \times 0.000005 \\ &= 1.884\ \Omega \end{aligned}$$

Inductive Reactance and Phase Angle

With all other factors held constant, changes in inductive reactance affect the impedance of the circuit. As the inductive reactance increases, the phase angle and the impedance increase. This is illustrated in Figure 9.

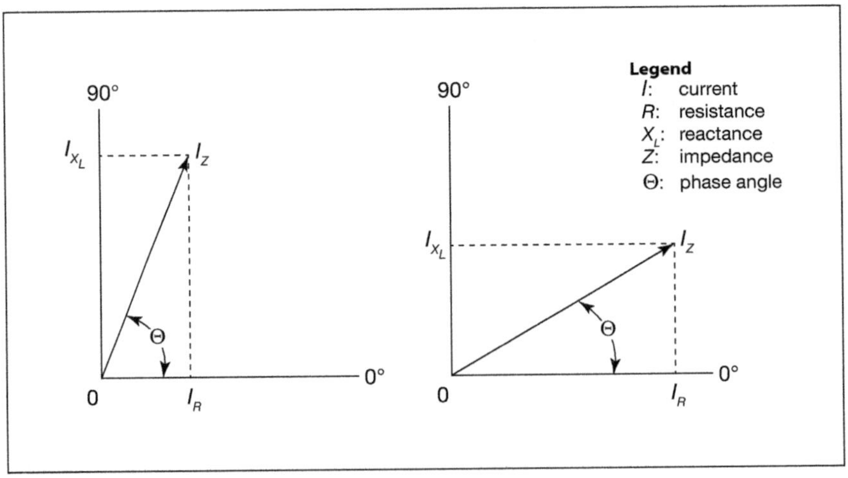

Figure 9. Current vector diagrams.

Impedance

In an AC circuit, impedance (symbolized by the letter Z) is the name given to the combination of resistance, capacitive reactance, and inductive reactance. For most testing applications, the capacitive reactance can be dropped from the equation since most eddy current probes have little or no capacitive reactance. Impedance in an AC circuit is the total opposition to the current flow through the circuit. The impedance unit is the ohm (Ω).

Since the resistance and the inductive reactance cause results that occur 90° out of phase with each other, they cannot be simply added together to determine the impedance. The simplest way to combine the resistance and the inductive reactance values to obtain the impedance value is through a vector diagram.

A vector is a line whose length represents its value and the direction represents its phase relationship. Figure 10a shows resistance and inductive reactance vectors 90° apart in direction. By adding these two vectors together, a rectangle can be constructed, and the diagonal from corner to corner represents the impedance (Z) and phase angle, as shown in Figure 10b.

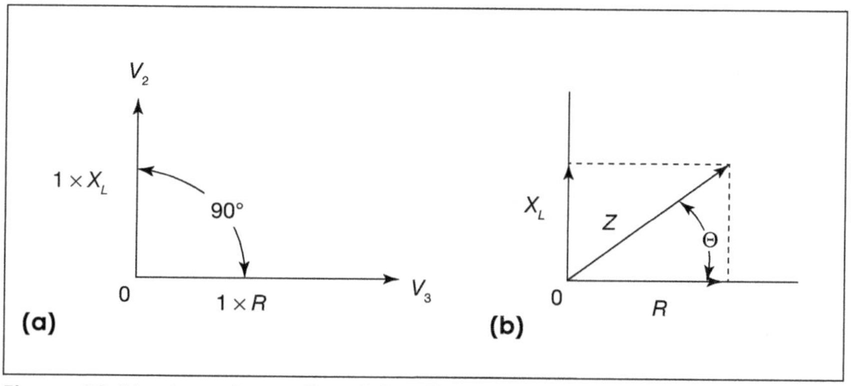

Figure 10. Vector schematics: (a) voltage-plane diagram; and (b) impedance-plane diagram.

The Greek philosopher Pythagoras developed what is known today as the pythagorean theorem, which states that in a right angle triangle, the square of the hypotenuse (side opposite the right angle) is equal to the sum of the squares of the other two sides. Using Figure 10b, it is determined that:

(Eq. 12a) $$Z^2 = R^2 + X_L^{\,2}$$

or

(Eq. 12b) $$Z = \sqrt{R^2 + X_L^{\,2}}$$

The concept that must be understood is what is meant by phase angle. Refer back to Figure 9. The phase angle between the resistance vector and the inductive reactance vector is 90°. The Greek letter theta (Θ) is used to denote the phase angle of the impedance vector.

Arctan

Tangent is defined as the ratio of the opposite side of a right angle triangle to the adjacent side. This is called the tangent of Θ.

(Eq. 13)
$$\tan\Theta = \frac{\text{length of opposite side}}{\text{length of adjacent side}}$$

The angle whose tangent is known is written $\tan^{-1}$ or arctan. Thus, the phase angle can be expressed as a function of the arctangent as follows.

(Eq. 14)
$$\Theta = \arctan\frac{X_L}{R}$$

Problem 5

An eddy current test is carried out at a test frequency of 60 kHz. The coil resistance is 10 Ω and its inductance is 5 μH. Calculate the following:

A. The inductive reactance of the test coil.
B. The impedance of the test coil.
C. The phase angle between the total impedance vector and the resistance vector.

Solution A

$$\begin{aligned} X_L &= 2\pi fL \\ &= \left[2\pi \times (60 \times 1000)\right] \times 0.000005 \\ &= (6.28 \times 60\ 000) \times 0.000005 \\ &= 376\ 800 \times 0.000005 \\ &= 1.884\ \Omega \end{aligned}$$

Solution B

$$\begin{aligned} Z &= \sqrt{R^2 + X_L{}^2} \\ &= \sqrt{10^2 + (1.884)^2} \\ &= \sqrt{100 + 3.55} \\ &= \sqrt{103.55} \\ &\approx 10\ \Omega \end{aligned}$$

Solution C

$$\begin{aligned} \Theta &= \arctan \frac{X_L}{R} \\ &= \arctan\left(\frac{1.884}{10}\right) \\ &= \arctan 0.1884 \\ &= 10.7^\circ \end{aligned}$$

>4

Magnetic Flux Leakage Theory

Introduction

Magnetism is a property possessed by certain materials by which they can exert a mechanical force on other similar materials. Magnetic flux leakage testing (MFL) is a nondestructive test method that can provide a quick assessment of the integrity of ferromagnetic material. This method involves magnetization of the test object by a permanent magnet or by passing an excitation current directly through an electromagnet. The presence of a discontinuity or thickness change on or near the surface of the material disturbs the magnetic flux lines and results in a local leakage field around the discontinuity.

The magnetic flux leakage can be detected using noncontact sensors, such as a hall effect probe or a simple induction coil. A hall effect probe uses an element oriented parallel to the sample surface that is sensitive to the normal component of the magnetic flux leakage field so as to generate a signal. (A typical signal is shown in Figure 1.) Essentially, the signal is a measure of the voltage difference across an electrical conductor (test object) transverse to an electric current in the conductor and a magnetic field perpendicular to the current.

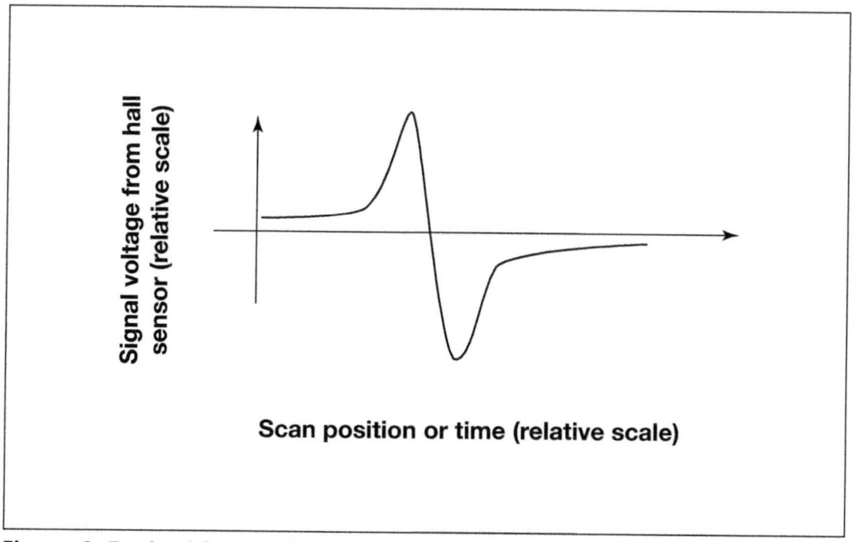

Figure 1. Typical flux leakage field signal for a rectangular notch in a test object.

A leakage field at an air gap in a longitudinally magnetized test object is shown in Figure 2. This disruption is produced by the alignment of magnetic domains in the stretched metal crack site. For this reason, cracks in the formation state are highly detectable and normally produce sharp, well-defined indications. Open cracks that have been subject to large thermal, chemical, or mechanical forces may have had their magnetic field disruption characteristics greatly or entirely reduced and may not be detectable using MFL.

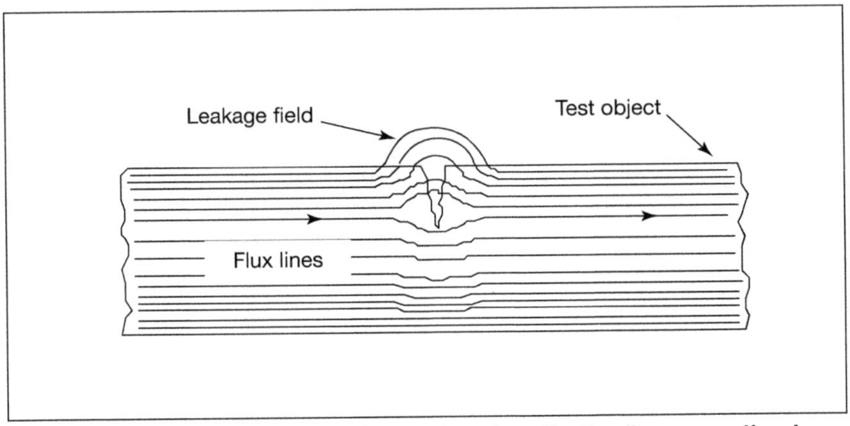

Figure 2. Leakage field at an air gap in a longitudinally magnetized test object.

Lines of Force

If a bar magnet is covered with a sheet of paper and iron filings are scattered over the paper, the filings align themselves along definite lines that pass from the poles of the magnet, as shown in Figure 3.

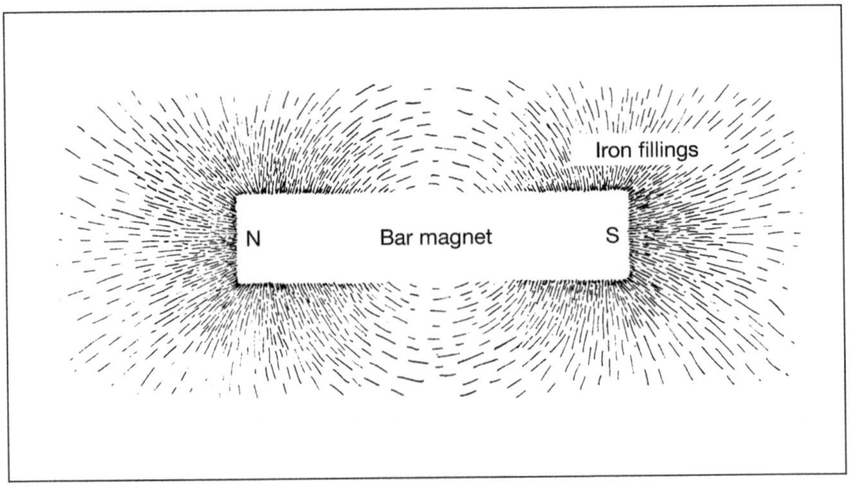

Figure 3. Magnetic field surrounding a bar magnet.

The alignment of the iron particles indicates that these lines form a field around the magnet, and any magnetizable material that enters this field is attracted to the magnet. For this reason the lines are called lines of force. Since the lines of force around a magnet seem to flow from the north to the south pole, they are often called flux lines. Flux is defined as a flow or flowing. Although lines of force are a useful concept theoretically, in actuality they do not exist!

Magnetic Poles

The regions close to the ends of bar or other magnets are known as magnetic poles, and these form a useful concept in magnetic theory. Lines of force are postulated to begin and end at magnetic poles.

Law of Magnetism

When like poles of magnets are brought together, the lines of force appear to repel each other, but when unlike poles are brought together, the magnets attract each other.

Magnetic Flux

Magnetic flux is the total amount of magnetism in a specific area. It is measured in webers. One weber (Wb) is equal to 1 volt-second (V·s).

(Eq. 1) $$1\ \text{Wb} = 1\ \text{V} \cdot \text{s}$$

Flux Density

The flux lines that surround a magnet are close together near a bar magnet. Figure 4 shows where the magnetic force is strongest, and it can be seen that the flux "flows" from south (S) to north (N) inside

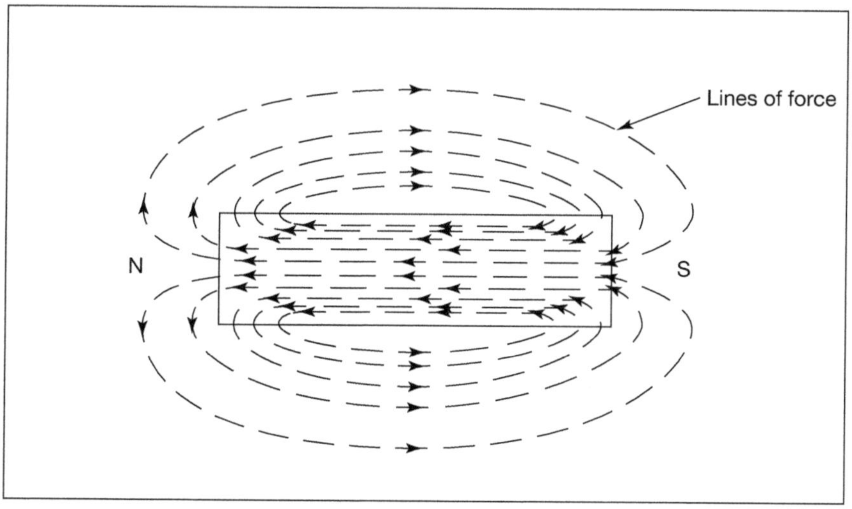

Figure 4. Distribution of flux around a magnet.

the bar magnet where it is quite dense, and from N to S outside the magnet. When spread farther apart away from the magnet, the magnetic field grows weaker. Thus, the density of the flux in a particular area determines the strength of the field in that area.

Conversely, the flux density can be determined by measuring the strength of the field. Flux density is defined as the number of lines of force expressed in webers of magnetism perpendicular to a specific area and is measured in webers per square meter (Wb/m^2). It can also be defined as one line of force at right angles to an area of 1 square centimeter (cm^2), in which case it is 1 gauss (G). There are 10 000 G in 1 Wb/m^2. The SI unit for magnetic flux density is the tesla (T). The conversion factors are as follows:

(Eq. 2a)
$$1\text{ T} = 10\ 000\text{ G}$$

or

(Eq. 2b)
$$\begin{aligned}1\text{ G} &= 10^{-4}\text{ T}\\ &= 0.0001\text{ T}\\ &= 0.1\text{ mT}\end{aligned}$$

Right-Hand Rule

As previously mentioned, it was discovered that when an electrical current flows through a wire, a magnetic field exists around the wire. The direction of the magnetic field around the wire depends on the direction of current flow through the wire. This relationship may be determined by the right-hand rule, as illustrated in Figure 5.

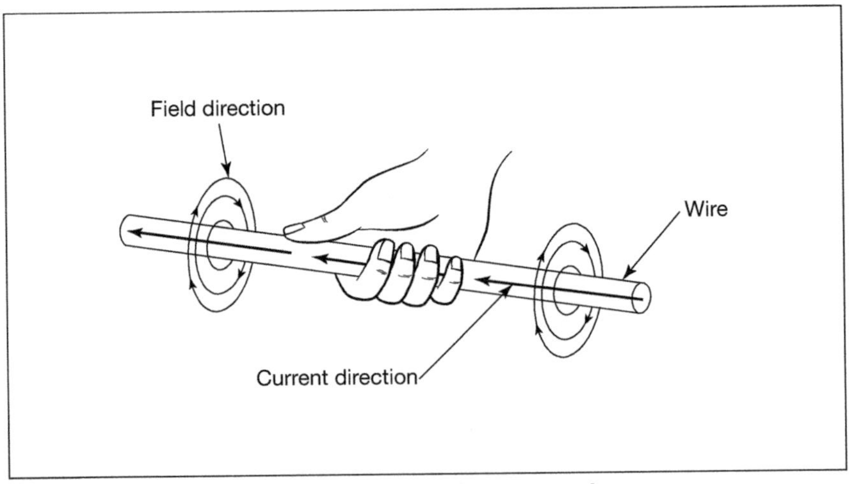

Figure 5. Magnetic field surrounding a bar magnet.

Magnetic Field Induced by Coil

If straight wire is wound into a coil, the lines of force encircling the wire form the magnetic field inside and outside of the coil, as illustrated in Figure 6. The field thus created is similar to the field of a bar magnet. The strength of the magnetic field is dependent upon two factors: the number of turns in the coil, and the magnitude of the current. Increasing either one increases the strength of the magnetic field.

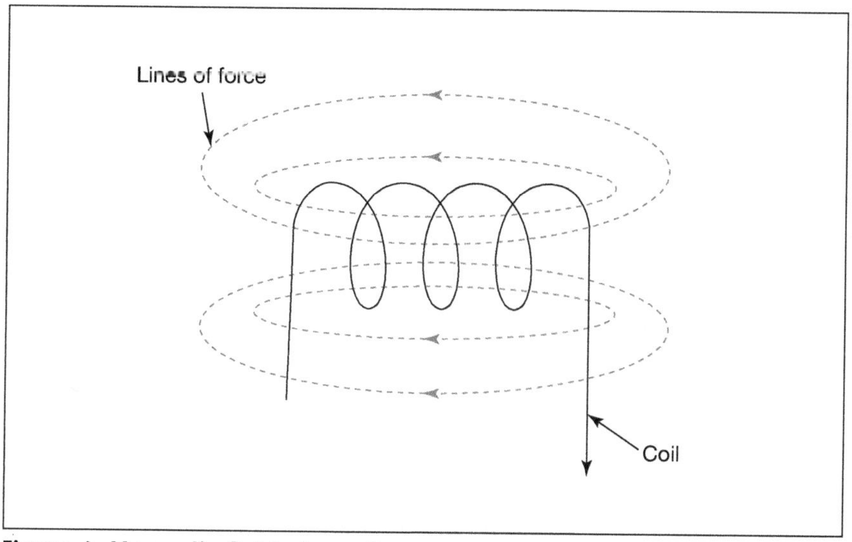

Figure 6. Magnetic field of a coil.

Electric current can be used to create a magnetic field in materials. If a coil is wrapped around a bar of iron and direct current passed through the coil, a magnetic field is established in and around the bar due to the magnetic field caused by the current flowing through the coil.

Magnetic Properties of Materials

Materials like iron that are magnetizable are called ferromagnetic materials. Ferromagnetic materials are capable of retaining some part of the magnetic field induced in them. Within the ferromagnetic group of materials, some are more easily magnetized than others. Those that are easily magnetized retain relatively little of the magnetic field after the current is shut off. Those that are difficult to magnetize retain more of the magnetic field after the current is shut off.

Magnetic Domains

Some materials that can be magnetized possess microscopic regions called domains that are fully magnetized (to magnetic saturation) in specific directions that are dominated by their crystal axis orientations and boundaries. These domains have a positive and negative polarity at opposite ends because of internal magnetic alignment. If the material is not magnetized, as shown in Figure 7a, the domains are randomly aligned, so the net magnetization appears to be zero from outside the material. Domain walls are defined by the edges of the microscopic crystallites or submicroscopic anomalies within the material.

When the material is subjected to an external magnetic field, some of the domains grow or shrink depending upon whether their axes are parallel or nonparallel to the external field strength. At higher external fields, the remaining domains begin to rotate into the field direction, as shown in Figure 7b, and many domains disappear as a result.

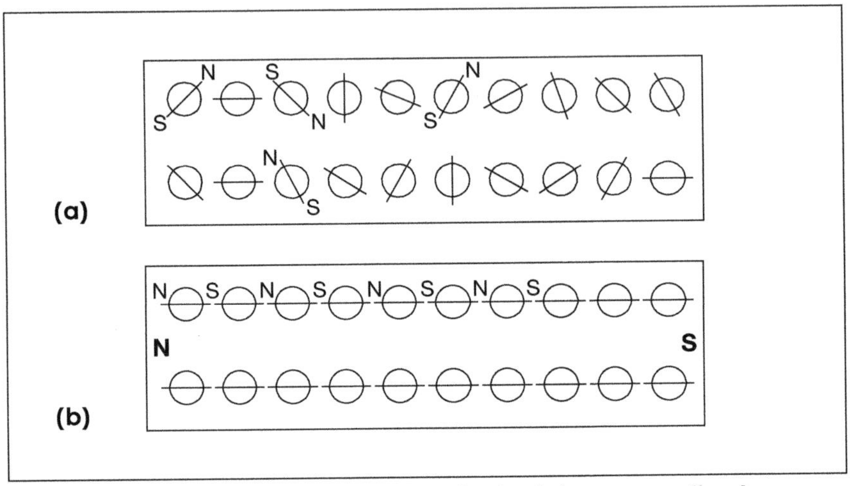

Figure 7. Alignment of magnetic domains in: (a) unmagnetized material; (b) magnetized material.

Magnetic Hysteresis

All ferromagnetic materials have certain magnetic properties that are specific to that material. Most of these properties are described by a magnetic hysteresis loop. The data for the hysteresis loop are collected by placing a bar of ferromagnetic material in a coil and applying an alternating current or direct current. By increasing the magnetizing field strength (H) in small increments, and measuring the flux density (B) at each increment, the relationship between magnetic field strength and flux density can be plotted. Thus, a hysteresis loop is also referred to as a B and H (B, H) curve.

The relationship between magnetic field strength and flux density is not linear for ferromagnetic materials. A specific change in H may produce a smaller or larger change in B, as shown in Figure 8, the initial curve for an unmagnetized piece of steel. Starting at point 0 (zero magnetic field strength and zero magnetic flux) and increasing H in small increments, the flux density in the material increases quite rapidly at first, and then generally slows until point A is reached. At point A, the material becomes magnetically saturated. Beyond the saturation point, increases in magnetic field strength do not increase the flux density in the material. In diagrams of full hysteresis loops, the curve 0-A is often drawn as a dashed line since it occurs only during the initial magnetization of an unmagnetized material. It is referred to as the virgin curve of the material.

When the magnetic field strength is reduced to zero—point *B* in Figure 8b—the flux density slowly decreases. It lags the field strength and does not reach zero. The amount of flux density remaining in the material (line 0-*B*) is called residual magnetism or remanence. The ability of ferromagnetic materials to retain a certain amount of magnetism is called retentivity.

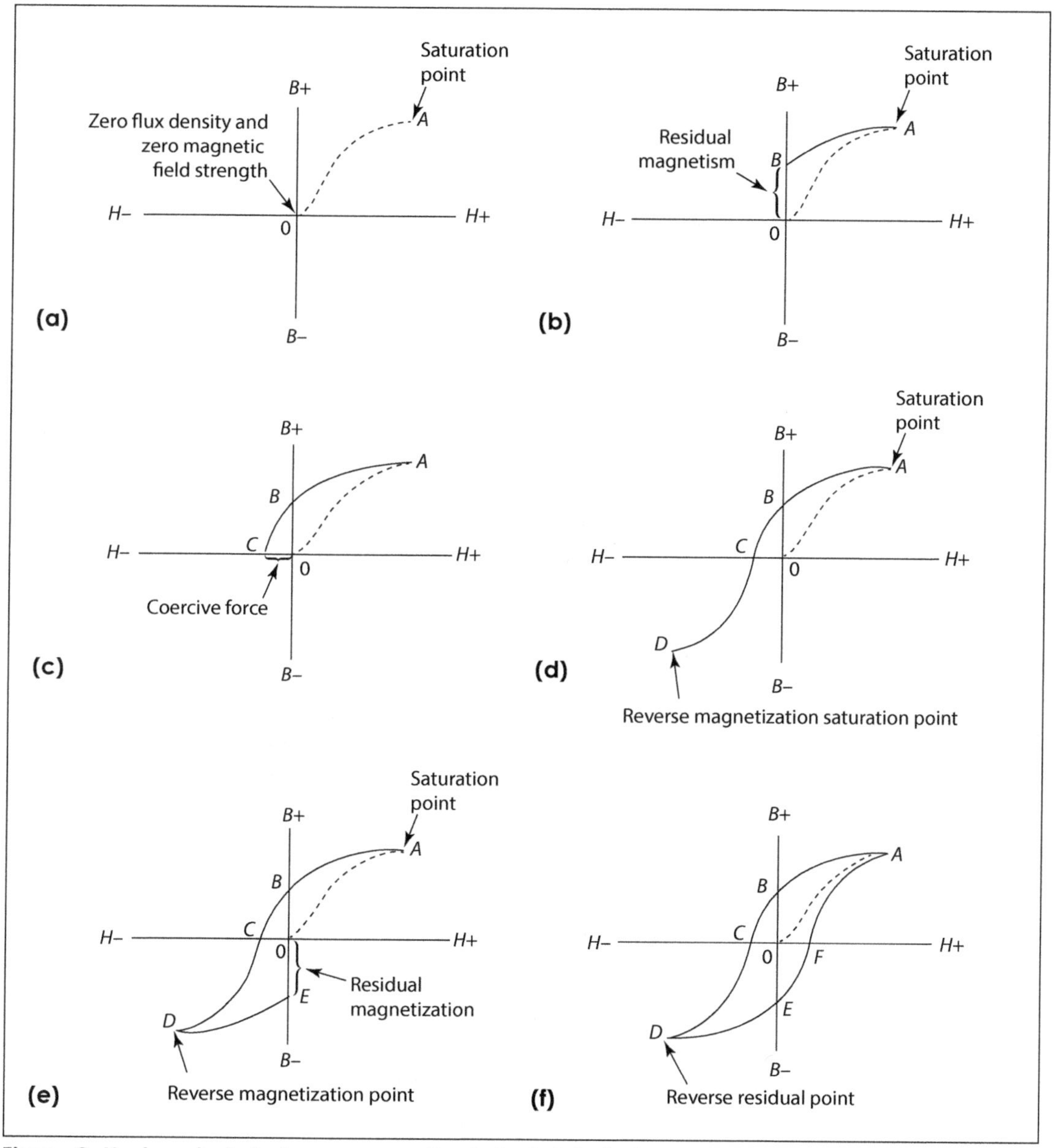

Figure 8. Hysteresis data for unmagnetized steel: (a) virgin curve of a hysteresis loop; (b) hysteresis loop showing residual magnetism; (c) hysteresis loop showing coercive force; (d) hysteresis loop showing reverse magnetism; (e) hysteresis loop showing reverse residual magnetism; and (f) complete hysteresis loop.

Removal of residual magnetism requires the application of a magnetic field strength in the opposite or negative direction, as shown in Figure 8c. When the magnetic field strength is first reversed and only a small amount is applied, the flux density slowly decreases. As additional reverse field strength is applied, the rate of reduction in flux density (line *B-C*) increases until it is almost a straight line (point *C*) where *B* equals zero. The reverse field strength required to reduce *B* to zero is called the coercive force.

If the amount of magnetic field strength is increased beyond point *C*, the magnetic flux changes its polarity and initially increases quite rapidly. It then gradually slows until point *D* is reached in Figure 8d. This is the reverse polarity saturation point, and additional magnetic field strength will not produce an increase in flux density. When the reversed magnetic field strength is reduced to zero—point *E* in Figure 8e—the flux densities of the residual magnetism from the straight and reversed polarities are equal (that is, line 0-*B* is equal to line 0-*E*).

Removal of the reversed polarity residual magnetism requires application of magnetic field strength in the original direction. Flux density drops to zero at point *F* in Figure 8f with the application of coercive force 0-*F*. Continuing to increase the field strength results in the magnetic polarity changing back to its original direction. This completes the hysteresis loop *A-B-C-D-E-F* (note that the curve *C-D-E-F* is a mirror image of curve *C-B-A-F*).

In terms of domains, what occurs during 0-*A* has been previously discussed. From *A* to *B*, some domains rotate back out of the direction of *H*. Further disordering occurs between *B* and *C* until, at *C*, the disorder is complete and the material is apparently demagnetized. The area between *C* and *D* reorients in the applied field direction. Thus, *D-E-F-A* is the reverse of what occurred along *A-B-C-D*.

Inner Hysteresis Loops

If point *A* in Figure 8 on the hysteresis curve is not reached during initial magnetization 0-*A*—that is, *H* is not strong enough to saturate the material with magnetic flux—an inner loop is obtained, which exhibits smaller values of residual magnetism at *B* and *E*, and lower values of the coercive force at *C* and *F*. Remanance and coercivity are the maximum values of the residual magnetism and coercive force, respectively.

Magnetic Permeability

One of the most important properties of magnetic materials is permeability. Permeability can be thought of as the ease with which materials can be magnetized. Air is assigned a permeability of one. Permeability (μ) is the ratio between the flux density and the magnetic field strength (B/H). It is also the rate of change of flux density (B) with respect to the magnetizing force (H), and it varies with position around the *B* and *H* curve.

(Eq. 3a) $$\mu = \frac{B}{H}$$

or

(Eq. 3b) $$B = \mu \times H$$

where

μ is permeability (pronounced "mu")
B is magnetic flux density
H is magnetic field intensity

Magnetic properties and hysteresis loops vary widely between materials and material conditions. They are affected by chemical compositions, microstructure, and grain size. Figure 9a is a hysteresis loop for hardened steel, and the loop is typical of a material with low permeability, high reluctance, high retentivity, and high residual magnetism that requires high coercive force for removal. Figure 9b is the hysteresis loop for an annealed low carbon steel. It is typical of a material with high permeability, low reluctance, low retentivity, and low residual magnetism that requires a low coercive force for removal.

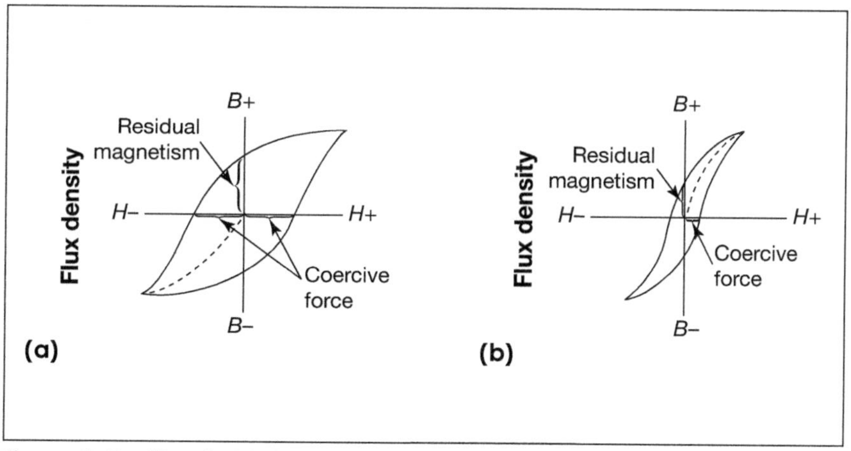

Figure 9. Positive field strength hysteresis loops: (a) hardened steel hysteresis loop; and (b) annealed low carbon steel hysteresis loop.

>5

Electromagnetic Testing Instrumentation

Introduction

In this chapter, you will find a general description of the instrumentation involved in electromagnetic testing, including eddy current, alternating current field measurement, and remote field testing. More specifically, you will learn about different types and arrangements of coils as well as the kinds of indicating instruments or reading apparatuses that are used in electromagnetic tests.

Eddy Current Instrument Circuits

There are many different types of eddy current instruments on the market today. They are similar in principle but vary in function and accessories. All eddy current instruments have some method of detecting the impedance or change of impedance in the test coil. It is the electronic circuitry that is the greatest variation between instruments.

A basic test circuit consists of an alternating current source supplying power to the testing coil. (See Figure 1.) A voltmeter is connected across the testing coil to measure the voltage. When the coil is placed on or near a test sample, the impedance of the coil changes. This change in impedance is reflected by the change in the reading of the meter.

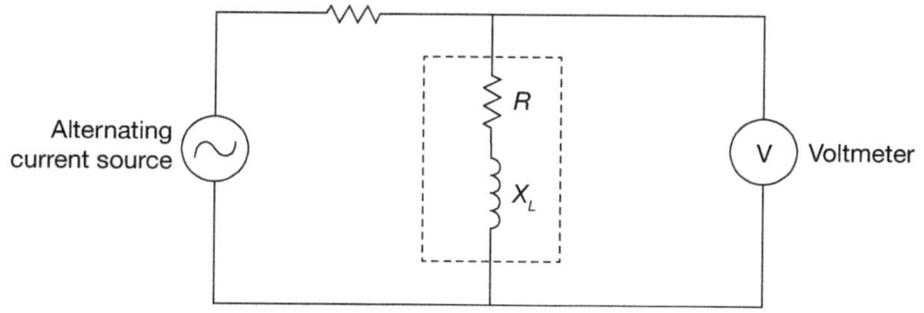

Figure 1. Basic test circuit.

Impedance Bridge Circuit (Wheatstone Bridge)

In a bridge circuit, also referred to as a wheatstone bridge, the alternating current source supplies power across the bridge that consists of two balanced resistors (R_1 and R_2), the testing coil, a balancing coil, and an ammeter to measure current (Figure 2). These units are connected in a bridge format: the resistor and the testing coil in one leg of the bridge, a resistor and a balancing coil in the other leg, and the ammeter across the two legs. When the bridge is in balance (that is, the impedance on both sides of the bridge is equal), the meter will read 0 A. When the testing coil is placed on or near a test sample, there is a change in impedance in that leg of the bridge. The bridge becomes unbalanced, and the ammeter indicates a current that is proportional to the imbalance. A differential amplifier can replace the bridge circuit.

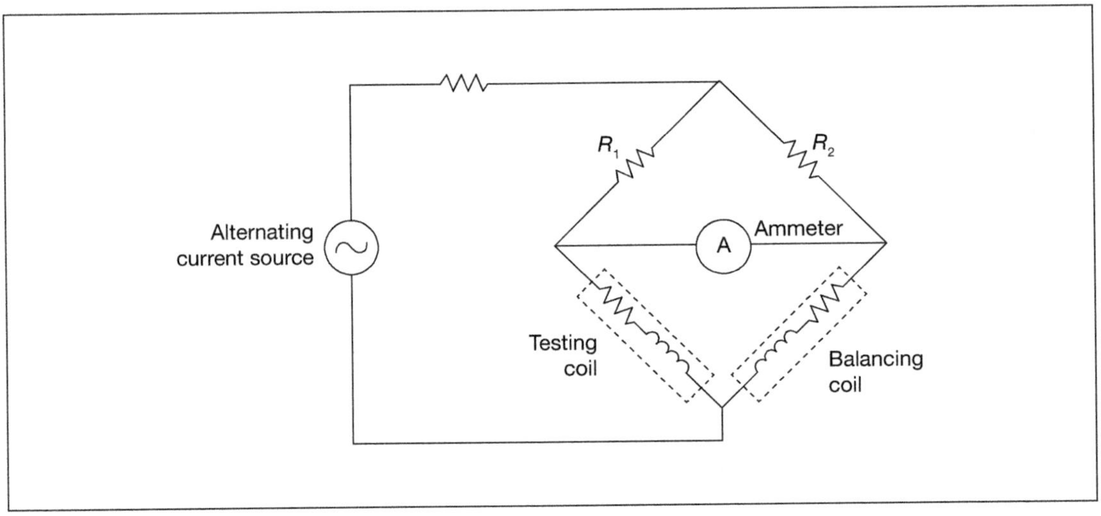

Figure 2. Basic bridge circuit.

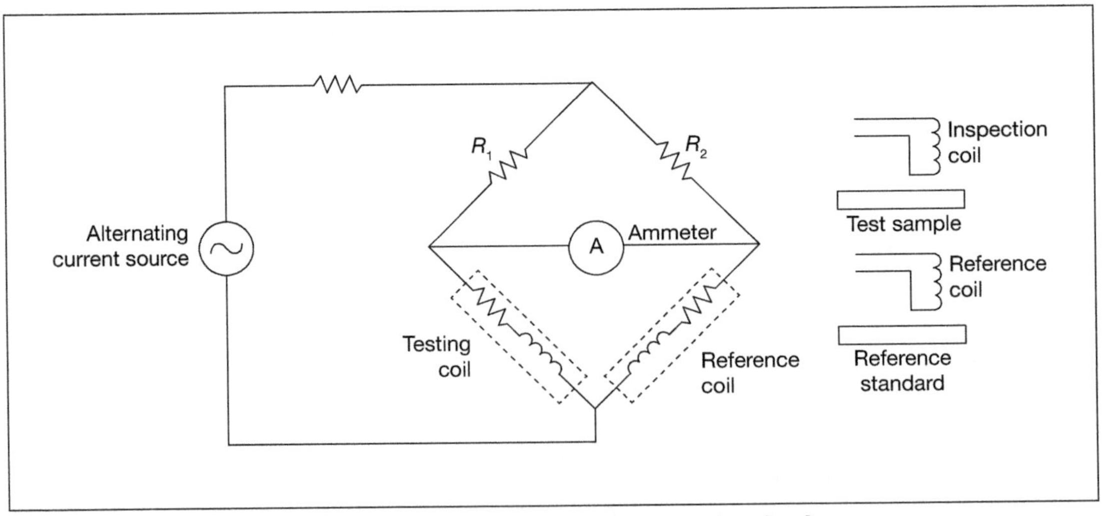

Figure 3. Bridge utilizing an inspection coil and a reference standard.

Another type of bridge circuit consists of two identical coils, one in each leg of the bridge (Figure 3). One of the coils is called a testing coil, and the other is called a reference coil. If the coils are identical, the bridge will be in balance. However, if the reference coil is placed on one sample of the material and the testing coil is placed on another sample of the material, the meter will show an imbalance only if there is some difference in the two samples.

Test Coil Arrangements

There are three basic coil arrangements:

- Absolute coil arrangement.
- Differential coil arrangement.
- Hybrid coil arrangement.

Absolute Arrangement (Single Coil)

Figure 4 illustrates the single coil arrangement. In this arrangement, the same coil is used to induce eddy currents in the test object and to sense the test object's reaction to the eddy currents.

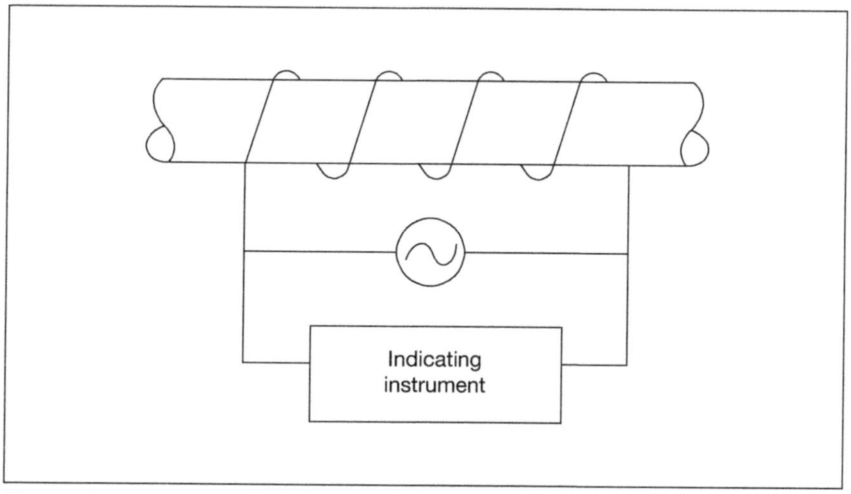

Figure 4. Absolute arrangement (single coil).

The single coil will test only the area under the coil and does not directly reference or compare itself to a reference standard. Because it tests the object without a comparison, it is called absolute.

Differential Coil (Self-Comparison Technique)

The differential coil arrangement shown in Figure 5 is a means of balancing out effects that are the same. The two coils (C_1 and C_2) are wound and electrically connected so that the output of one coil cancels the output of the other coil (that is, the outputs oppose each other). This occurs when the test object properties are the same under both coils. In other words, so long as there is no difference in material properties

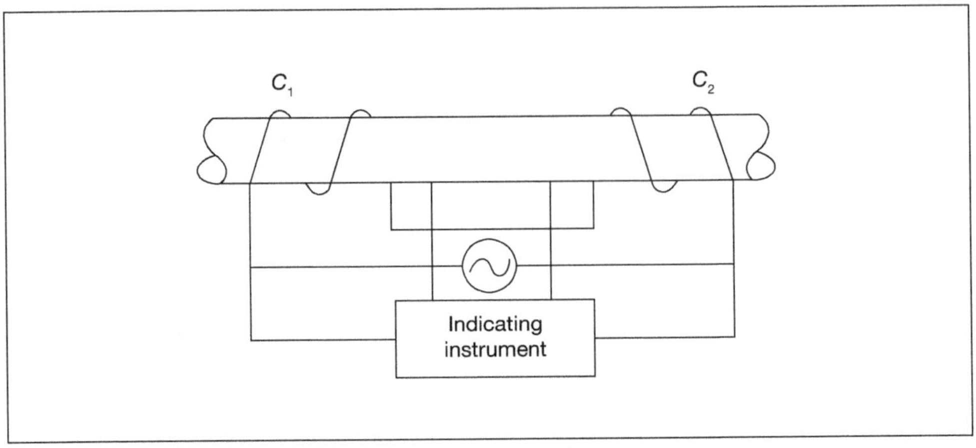

Figure 5. Differential coil (self-comparison technique).

under both coils, there is no indication on the output device. However, when a discontinuity or inhomogeneity is located under either one of the coils, an imbalance occurs. The differential coil arrangement is known as the self-comparison technique.

The self-comparison technique is insensitive to test object variables that occur gradually. Variables such as gradually changing wall thickness, diameter, temperature, or conductivity that affect both coils at the same time are not detected with the self-comparison differential coil.

Differential Coil (External Comparison Technique)

The coil arrangement illustrated in Figure 6 is exactly the same as the self-comparison technique, except that the differential coil arrangement is set up on an external reference standard and the test object. A carefully chosen, discontinuity-free test reference object is held stationary in one coil while the test object moves through the other coil.

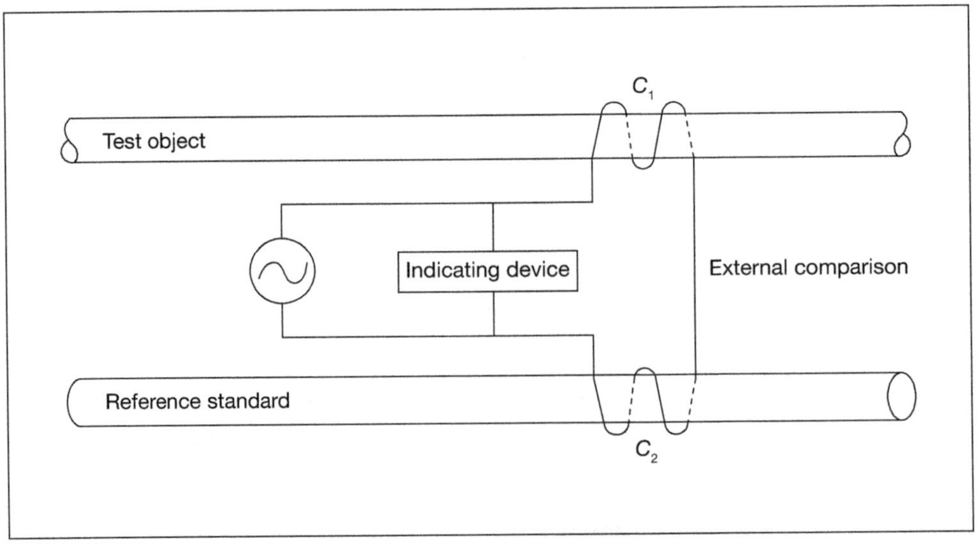

Figure 6. Differential arrangement (external comparison technique).

Here, coil C_2 and the discontinuity-free test object are set up as a reference standard. As the test object passes through coil C_1, a comparison is made with the reference standard. No indication is observed unless a discontinuity or other change such as thickness, permeability, or conductivity appears in the test object. If a discontinuity passes through coil C_1, the output of the coils become unbalanced and an indication is obtained.

This differential coil arrangement technique is used to detect differences between a reference standard and test object. It is mostly used to compare conductivity, permeability, and dimensional measurements. The problem with this technique is the large number of reference standards required and the tolerances that exist in diameter and metallurgical composition.

Hybrid Coil Arrangements

Hybrid coils may or may not be the same size and are not necessarily adjacent to each other. Common types of the hybrid coil are driver/pickup, through-transmission, and primary/secondary coil assemblies. The through-transmission technique involves inducing eddy currents into the test object by a transmitting or driver coil placed on one side of the material. The presence of eddy currents is sensed by a receiving or pickup coil placed on the opposite side of the material, as shown in Figure 7. This arrangement requires that the two coils be placed exactly opposite each other. The voltage in the sensing coil is a function of the current magnitude and frequency applied to the excitation coil, coil parameters of the exciting and sensing coils, and the test object characteristics. Only thin materials may be tested with through-transmission coils.

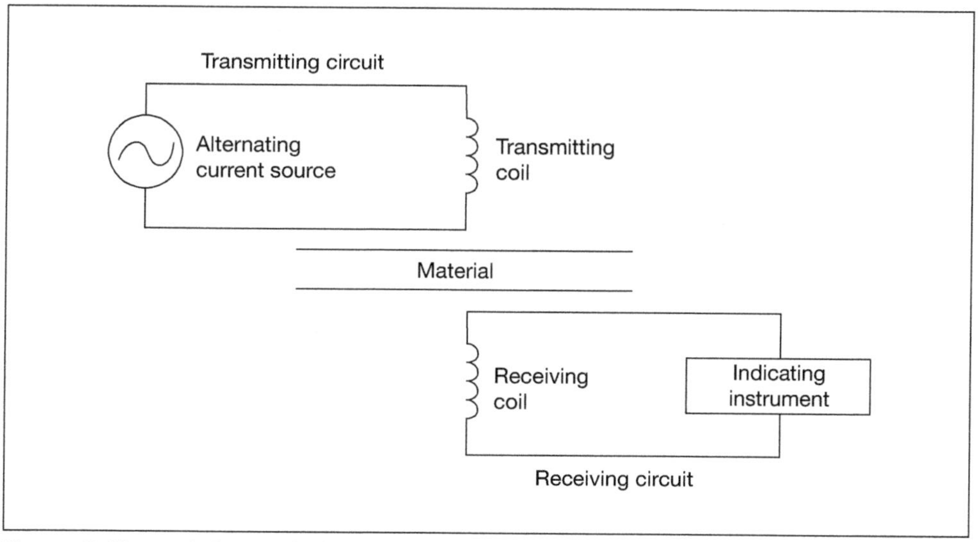

Figure 7. Through-transmission arrangement.

There are many other hybrid coils where both the driver coil and pickup coil (or coils) are in the same housing. These cannot be called through-transmission, but they operate the same way. They are typically designed and built for very specific test geometries and are an attempt to overcome some of the natural limitations of surface probes. In the aerospace industry, these are common for the detection of subsurface cracks and corrosion.

Internal Functions of Eddy Current Instrumentation

For an eddy current system to provide information to a technician, five functional steps have to be performed, as shown in Figure 8:

1. Signal excitation.
2. Signal modulation.
3. Signal preparation.
4. Signal demodulation.
5. Signal display.

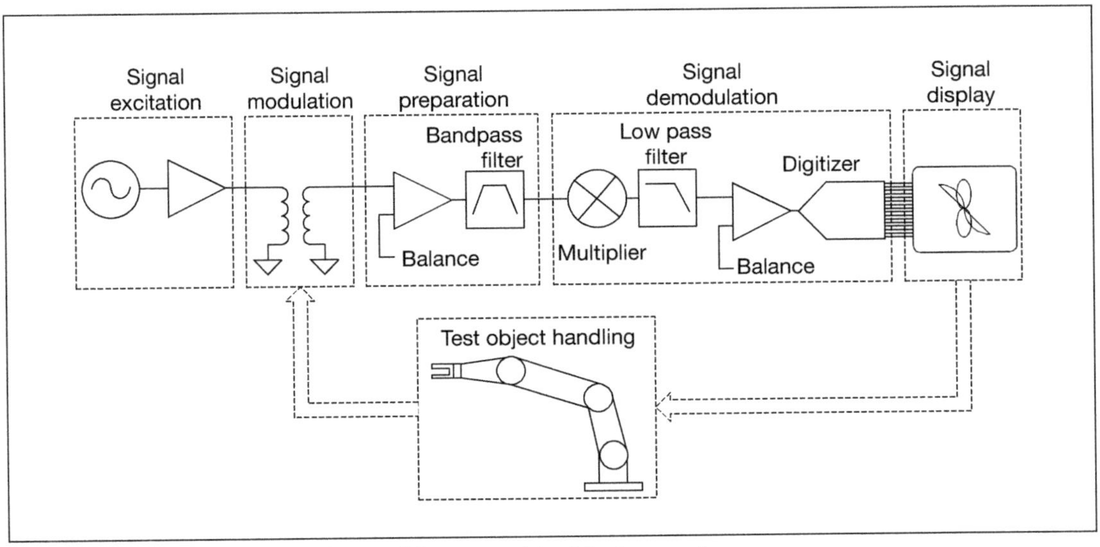

Figure 8. Internal function of an eddy current test instrument.

Signal Excitation

The excitation portion of an eddy current instrument consists of a signal generator and amplifiers to drive the test coils. The signal generator (or oscillator) provides sine wave excitation for the test coil. Single-frequency systems have one fixed frequency, whereas multifrequency systems can apply several frequencies to provide multiple parameter options. The application determines the required frequency and the number of frequencies to be used.

Signal Modulation

Signal modulation occurs in the electromagnetic field of the coil's assembly. It is the magnetic field created by the primary coil that provides the energy transfer into the test object. This magnetic energy is modulated by the test object, and the resultant magnetic field from the eddy currents opposes the primary field and is sensed by the instrument for processing.

Signal Preparation

After modulation, the signal is processed for demodulation and analysis. The purpose of this step is to amplify the probe signal and reject extraneous noise. This part of the instrument may consist of a single-ended amplifier in an absolute or simple driver pickup system, differential amplifiers for a differential bridge, or a more sophisticated system of driver and pickup. The bandpass filter or speed filter is selected to detect discontinuities at a given inspection speed.

Signal Demodulation and Analysis

In this step and after the signal has been demodulated, the signal can be analyzed by several means, including digital meters or digital displays for further manipulation and analysis.

Signal Display

The signal display section is the key link between the test equipment and its intended purpose. The signal can be displayed in many different ways. Common displays include liquid crystal displays (LCDs) and computer screens.

Signal-to-Noise Ratio

The signal-to-noise ratio (SNR) is the ratio of signals of interest to unwanted signals. Common noise sources are test object variations of surface roughness, geometry, and homogeneity. Other electrical noises can be caused by external sources, such as welding machines, electric motors, and generators. Mechanical vibrations can increase test system noise by physical movement of the test coil or test object. In other words, anything that interferes with a test system's ability to define a measurement is considered noise.

Improving Signal-to-Noise Ratio

The SNR can be improved by several methods. If a test object is dirty or scaly, the SNR can be improved by cleaning the test object via shot blasting, pickling in acid, wire brushing, or light abrasive belts. Electrical interference can be shielded or isolated. Phase discrimination and filtering can also improve the SNR.

It is common practice in nondestructive testing to require a minimum SNR of 3:1. This means that a signal of interest must have a response at least 3× that of the noise at that point.

The absolute noise level and the absolute strength of a signal from a discontinuity depend on several factors, such as the search coil type and size, the frequency, the inspection path and distance, and the test object surface condition and microstructure. Other factors include the discontinuity size, location, and orientation.

Alternating Current Field Measurement Instrumentation

The alternating current field measurement technique involves inducing a locally uniform current into a test object and measuring the magnetic flux density above the test object surface. The presence of a surface-breaking discontinuity perturbs the induced current and the magnetic flux density. Relative, rather than absolute, amplitudes of components of the magnetic flux density are used to minimize variations caused by material properties, instrument standardization, and other factors.

These relative amplitudes are compared with values in sizing tables produced from a mathematical model to estimate discontinuity sizes without the need for standardization using artificial discontinuities such as slots in a reference block. The sizing tables have been produced by repeated running of the model for semielliptical cracks in a wide range of different lengths and depths.

Software is used to control the instrument and to display and analyze the data. The required locally uniform magnetic field is induced using one or more horizontal axis solenoids, with or without a yoke. (See Figure 9.)

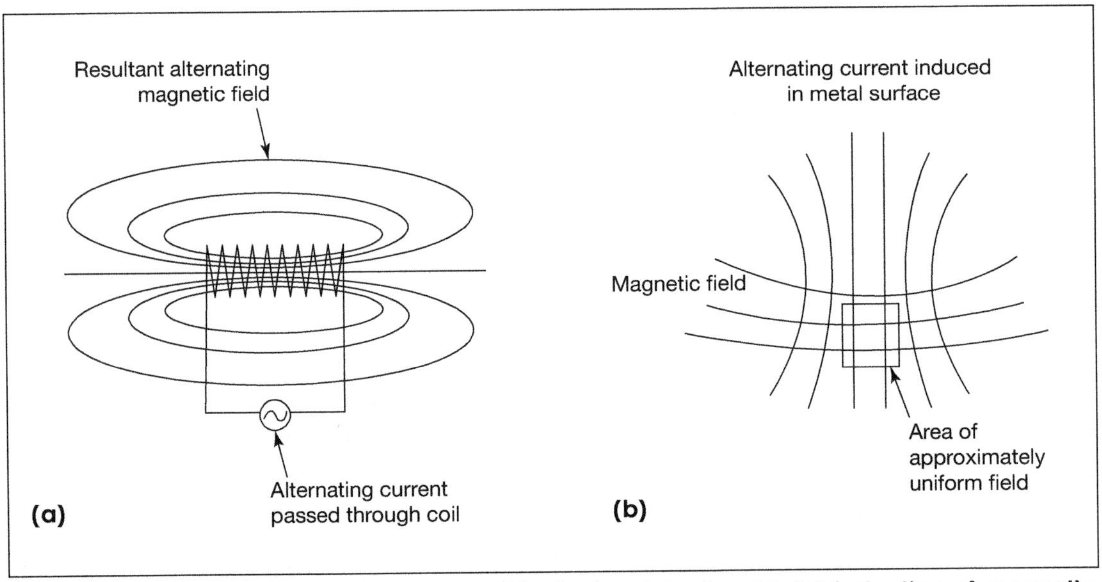

Figure 9. Uniform magnetic field generated by horizontal solenoid: (a) induction of magnetic field by alternating current in coil; (b) uniform field induced by alternating current in metal surface.

By convention, the direction of this electric field (E) is designated as the Y axis and the direction of the associated uniform magnetic flux density (B) (at right angles to the electric field and parallel to the test surface) is designated as the X axis. The Z axis is then the direction normal (perpendicular) to the surface. (See Figure 10.)

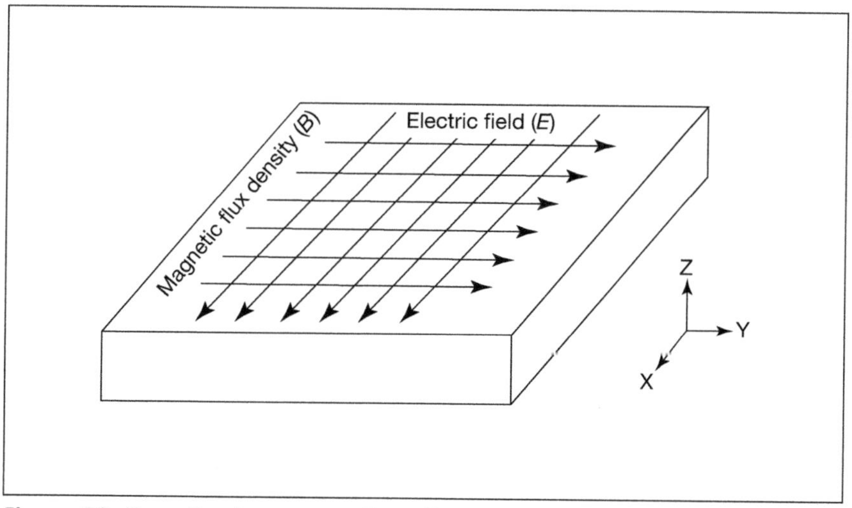

Figure 10. Coordinates conventionally used in alternating current field measurement.

Measurements of B in the X (B_x) and Z (B_z) directions from magnetic sensors in a handheld probe are used with software algorithms to determine the length and depth of the discontinuity.

The actual parameters used by the software can vary but must include the following:

1. The perturbation amplitude is needed for one component of the magnetic flux density produced by the discontinuity (usually B_x, but B_z can also be used).
2. The intensity of the input magnetic flux density B_x is used to normalize the perturbation. This background B_x value must therefore be measured in an area of properties similar to the perturbation value. This area is normally next to the discontinuity but outside its influence.
3. A measurement unit is needed to quantify the signal. This unit is usually the distance between the peak and trough in the B_z signal because these signals are sharply defined, but the distance between the peaks in B_x can also be used.

Remote Field Testing Instrumentation

Besides the coils and coaxial cable, a remote field testing system contains four other major components:

1. An oscillator is used as the signal source for the exciter coil and as a reference for the detector signal.
2. A power amplifier increases the power level from the oscillator signal so that it can be used to drive the exciter coil.
3. The phase and amplitude detector measures the detector coil signal.
4. A microcomputer-based storage device processes and stores the data.

Figure 11 shows how the different electronic components interact.

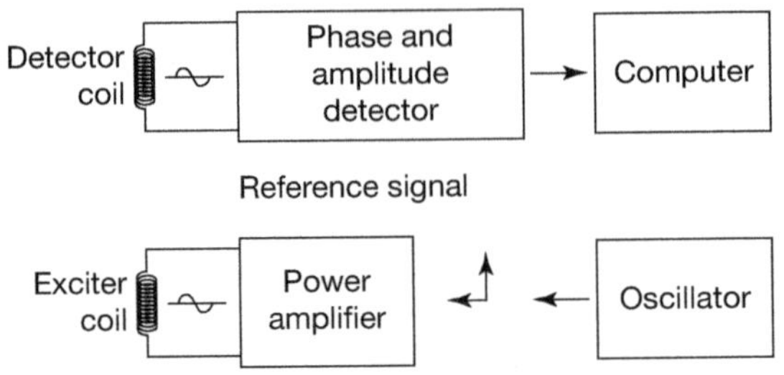

Figure 11. Electronic components of a remote field test system.

Driving of Remote Field Test Probe

The exciter coil is energized with alternating current at frequencies ranging from 50 Hz to 1 kHz for ferrous materials. Higher frequencies are used for tubes with lower permeability values. The exciter coil typically carries a current of 0.1 to 1.0 A, the limitation being probe temperature and, for some probes, magnetic saturation of the probe core. The detector signal depends directly on the exciter current and frequency. Thicker wall penetration can be obtained by lowering the frequency and increasing the exciter current. The actual test parameters, such as test frequency, drive voltage, and sample rate, are chosen by taking into consideration factors such as probe pull speed, discontinuity sensitivity requirements, magnetic permeability (μ), tube electrical conductivity (σ), and tube wall thickness (τ).

In general, a lower frequency, up to 250 Hz, is used for thick-walled and high-permeability pipe. For superior sensitivity or high test speeds, a higher frequency is preferred. The frequency is chosen as high as possible while minimizing noise and remaining in the remote field zone.

Because remote field testing is used in very diverse applications, it is important to check for the presence of any electromagnetic interference and noise sources, such as welders, electric motors, power inverters, and pumps. Such devices tend to generate noise in the frequency range of remote field testing. Remote field testing equipment manufacturers sometimes provide the user with a noise spectrum, showing the environmental noise for a range of frequencies. Some technicians prefer to place the probe in a thick-walled block to determine the baseline noise at a given frequency.

Because remote field testing signals are often quite small (1 to 10 μV), it is advisable to avoid using line voltages and their harmonics (60 Hz or, in Europe, 50 Hz), which can cause interference. In this case, the inspector needs to minimize the background noise while keeping in mind the aforementioned test factors.

Readout Mechanisms

An important part of an electromagnetic test system is the part of the instrument that quantifies the change in impedance. There are several different types of devices used. The device may be an integral part of the test set, it may be a module that is plugged into the test set, or it may be a separate unit connected to the test set with a cable. The indicating device used should be of adequate speed, accuracy, and range to meet the requirements of the test system. Permanent test records may be required by inspection codes so that corrosion or discontinuity growth rates, for example, can be monitored.

Analog Meters

An analog meter is an indicating device whose visual output varies as a continuous function of the input to the meter. A meter has a needle that moves in response to the input. The response is immediate, and scales can usually be calibrated to read specific values directly. The displays of analog instruments are relatively simple, enabling the user to view and perform simple manipulations of raw data. Generally, controls for phase, gain, alarm levels, and some filtering are available. Analog recording of data via magnetic tape and strip chart recorders were common in the 20th century but have largely been replaced by digital data storage. However, analog instruments may still be used in a few niche applications.

Audio Alarms

Audio alarms only indicate an abnormal condition. Alarm lights and audio alarms are commonly incorporated in electromagnetic test equipment. The indicator light and audio alarm give only qualitative information about the item, whether or not a condition is present.

Strip Chart Recorders

A strip chart recorder is a device that records an electrical or mechanical input trend onto a piece of paper (analog) or on a computer screen (digital). As previously noted, digital recording devices are replacing analog strip charts. In general, the same principles apply. Strip chart recorders provide a recording of values at reasonably high speeds. Several channels can be recorded at the same time. Strip chart recordings are common in testing tubing where the discontinuity's location down the tube is critical. The strip chart length is indexed to time or distance and indicates normal or abnormal conditions. An example of a strip chart recording compared with a phase amplitude diagram is shown in Figure 12 for signals from differential and absolute coils in a remote field test.

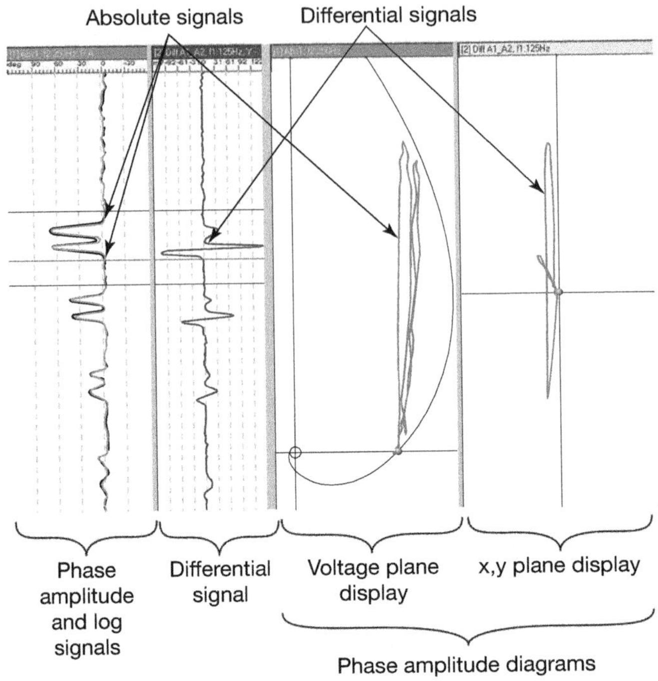

Figure 12. Strip charts and phase amplitude diagrams showing indications from a short discontinuity in a pipe wall.

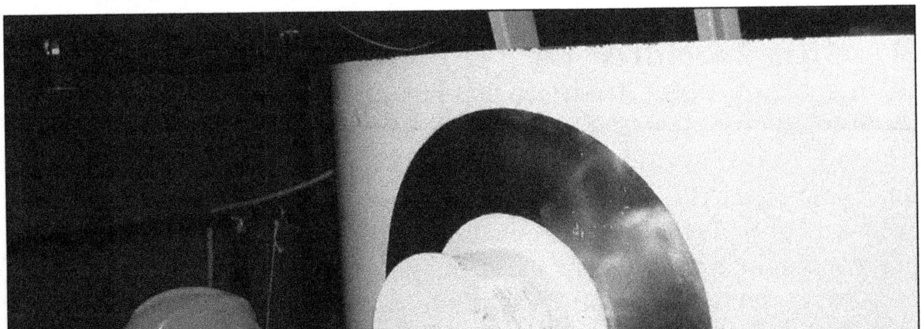

Figure 13. Eddy current test (using a digital display) of a port weld in a nuclear reactor storage cask for spent fuel bundles.

Digital Meters

A digital meter is one whose visual output is shown in discrete steps in time. The meter measures the input at a given moment, and the value of the measurement is displayed in numerical form. Since the readout is in numbers, the chance of technician error is less than when analog meters are used, but the output is relatively slow. Digital meters provide greater accuracy and range than analog meters. Digital conductivity meters, standardized from conductivity reference standards, feed acquired data into algorithms that calculate conductivity and then display the data in a numerical format, as shown in Figure 13.

Digital Displays

Digital displays of the output of a test circuit give an instantaneous, continuous presentation; are highly accurate; provide calibration capabilities so that values may be read directly; and have a broad range. In addition, the presentation is adjustable and stored so that parameters of particular interest may be studied more closely. Very few analog cathode ray tubes (CRTs) remain in service. Modern test systems employ digital screens that provide the same type of visual responses but with much greater flexibility in processing and manipulating the images. Systems that run from a software package loaded on a PC still use an X-Y CRT-type format to display data during data acquisition and for later data analysis.

Digital Data Storage

Digital data are generally displayed in a complex plane presentation with supporting strip chart and C-scan displays, as required by the application. The point described by the in-phase and quadrature (X and Y) components of the signal is displayed as a flying dot, as diagrammed in Figure 14, and the digital capabilities of the instrument allow variable persistence, centering of the dot, rotation of the signal, and scaling

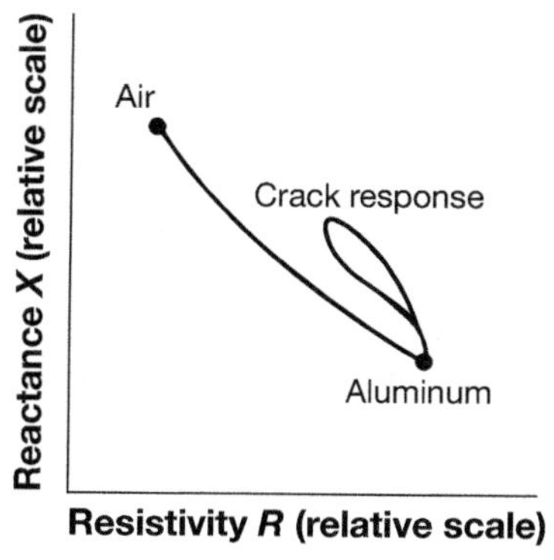

Figure 14. Tracing of the "flying dot" impedance plane response from a crack in an aluminum sample; the liftoff response curve connecting air to aluminum is also shown.

of the display. Digital systems facilitate the setup of standardization curves constructed from stored data and automated analysis of signals as compared to these curves. (Standardization curves are discussed in greater detail in Chapter 9.)

Digital Mixing

The combination of components from different test frequencies allows the suppression of unwanted parameters or signals from structures, such as support plates in tubing applications, while retaining the signature of discontinuities beneath those structures.

Alarms can be constructed digitally in any way—for example, as amplitude levels, boxes, or ellipses—for rejection or acceptance of test objects. Alarms from various frequencies or coils can be tagged to allow discrimination of different test object conditions. (See Figure 15.)

(a)

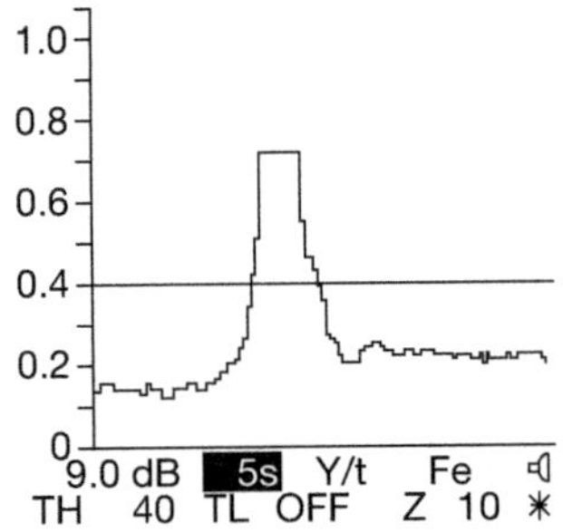

(b)

Figure 15. Electromagnetic test system with amplitude set for accept/reject of signal teeth from gear assembly: (a) signal tooth with large crack; (b) printout of computer display showing indication from the discontinuity in Figure 15a.

Liquid Crystal Display

A liquid crystal display (LCD) is a thin, flat display device made up of any number of color or monochrome pixels arrayed in front of a light source or reflector. It uses very small amounts of electric power and is suitable for battery-powered electronic devices.

Important factors to consider when evaluating an LCD monitor include resolution, viewable size, response time, matrix type, viewing angle, color luminosity, contrast ratio, and input ports (for example, DVI, VGA, or HDMI). A computer screen is a typical LCD that is widely used in state-of-the-art electromagnetic test instruments, as shown in Figure 16.

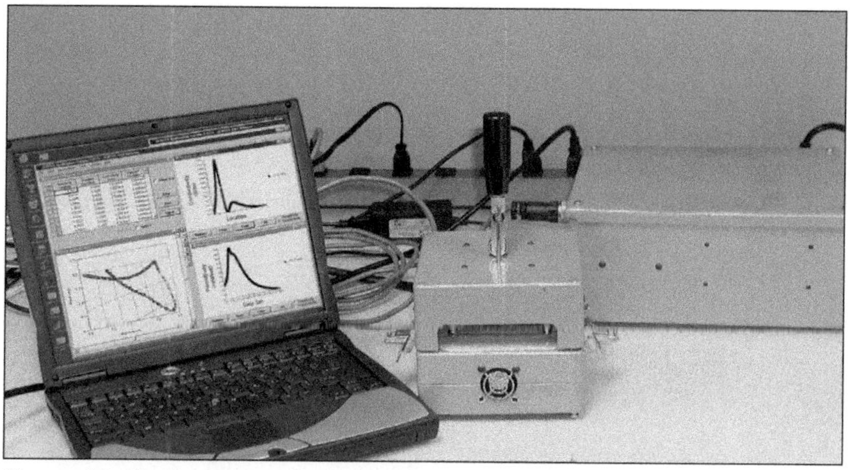

Figure 16. Electromagnetic test system including impedance measurement instrumentation, probe electronics, and interchangeable sensors for characterizing insulating (dielectric) materials.

6

Electromagnetic Sensing Elements

Eddy Current Inspection Coils

Inspection coils for eddy current testing are available in a variety of forms and can be arranged in a variety of ways. The choice of the arrangement of the inspection coil depends on the test situation. There are three types of inspection coils related to their physical structure, arrangements, and testing they perform:

- surface coil
- internal coil
- encircling coil

Surface Coil

Surface coil, probe coil, flat coil, and pancake coil are all common terms used to describe the same type of test coil used to examine the surface of a test object.

Figure 1 illustrates typical surface probes. Figure 1a shows the coil mounted in a probe, while Figure 1b shows the coil mounted in a spring-loaded housing. The spring loading ensures that a constant pressure is applied at all times to prevent separation (liftoff) of the coil from the

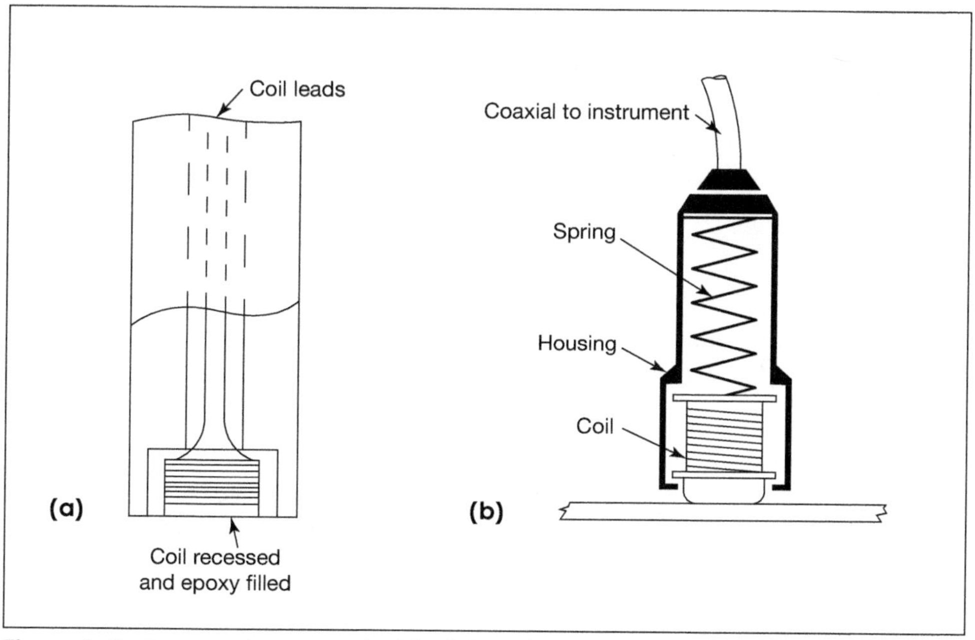

Figure 1. Surface probes mounted in: (a) probe; (b) spring-loaded housing.

surface of the test object. The surface probe may be handheld or may be mounted in automated scanning equipment. The coil, mounted in the end of the probe, is provided with a protective coating to serve as a wear surface and to standardize liftoff. The magnetic field produced by the coil is approximately the size of the coil.

Probe coils and probe coil forms can be shaped to fit particular geometries to solve complex testing problems. As an example, probe coils fabricated in a pencil shape, referred to as pencil probes, are typically used to test threaded areas of mounting studs and nuts or serrated areas of turbine wheels and turbine blade assemblies. Probe coils with added coil shielding may be used where high resolution is required.

A variation of the surface probe is shown in Figure 2. The coil is mounted in a holder shaped and sized to serve a special function. Holders may be designed to serve any particular requirement (such as the round surface of a tube or bar) or to position the coil at a particular place (such as the leading edge of turbine blades). Spring loading ensures that constant pressure can be maintained and prevents separation or liftoff.

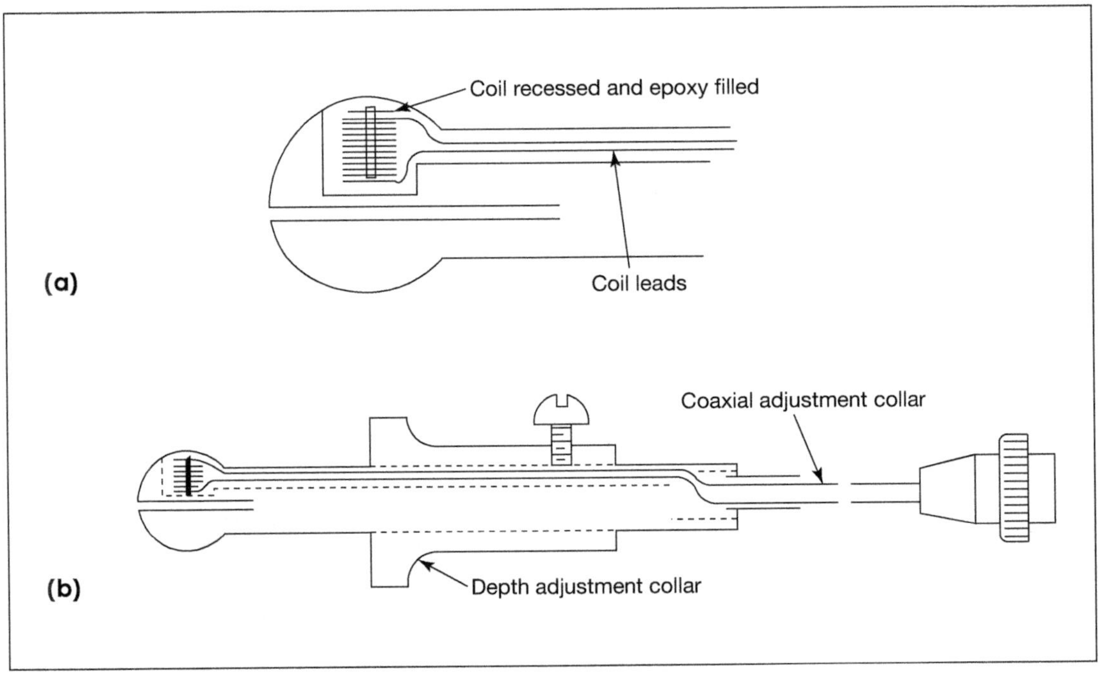

Figure 2. Hole probe: (a) close-up of coil; (b) schematic.

Applications

When using a high-resolution probe coil, the test object surface must be carefully scanned to ensure complete test coverage. This careful scanning is very time consuming. For this reason, probe coil tests of large objects are usually limited to critical areas. Probe coils are used extensively in aircraft testing for crack detection near fasteners and fastener holes. In the case of testing fastener holes (for example, bolt holes or rivet holes) and critical areas of heat exchanger tubes, the probe coil is spinning while being withdrawn at a uniform rate. This provides a helical scan of the hole using what is referred to as a spinning probe technique.

Internal Coil

Bobbin coil, inside diameter coil, and inside probe are terms that describe coils used to test from the inside diameter or bore of a tubular test object.

Figure 3 illustrates a type of coil that can be inserted into tubing to test for discontinuities and thickness changes in the tube. The internal coil induces currents that encircle the entire circumference of the tube so that the entire section surrounding the coil is tested.

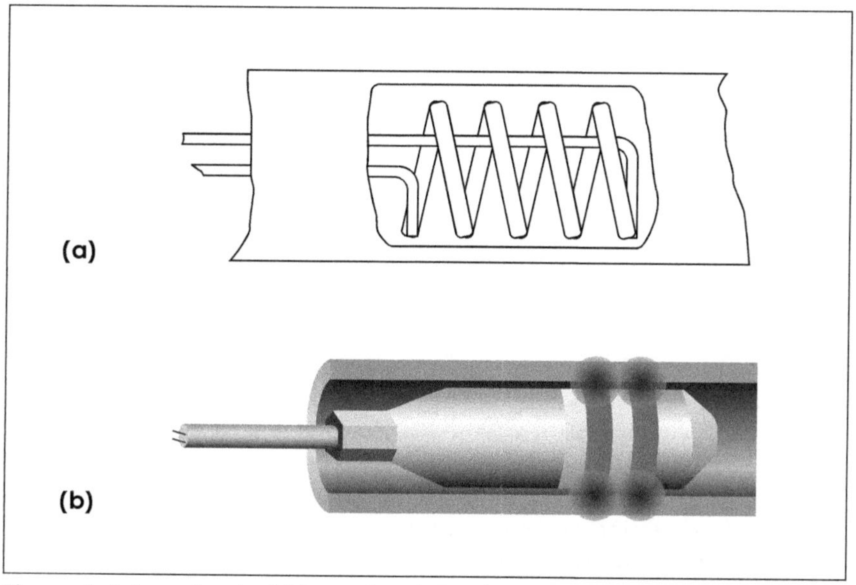

Figure 3. Internal or bobbin probe: (a) schematic; (b) cutaway view.

Because the currents induced in the material are strongest near the coil, the internal coil is more sensitive to discontinuities on or near the inner surface of tubing, while external coils are more sensitive to discontinuities on or near the outer surface. The internal coil may be either wide or narrow, and the magnetic field extends slightly beyond the ends of the coil. The coil also may be shielded or unshielded.

Applications

As noted, internal coils are primarily used to test tubular products from the inside diameter or the bore of a tubular test object, for example, tubes examined in heat exchanger applications.

Encircling Coil

Encircling coil, outside diameter coil, and feed-through coil are common terms to describe coils that are used primarily to test outside diameter surfaces of objects that pass through the coil. Although the coil is the same in both encircling and inside diameter probes, the field distribution is somewhat different. The flux density gradient tends to be more uniform inside the coil and decreases to zero at the center of the tested material.

The width of the coil (Figure 4) is a function of the application. Wide coils cover large areas, so they respond mostly to bulk effects—that is, conductivity—whereas narrow coils sense small areas and so are more responsive to small changes produced by discontinuities or small thickness changes. The magnetic field of the coil extends beyond the ends of the coil.

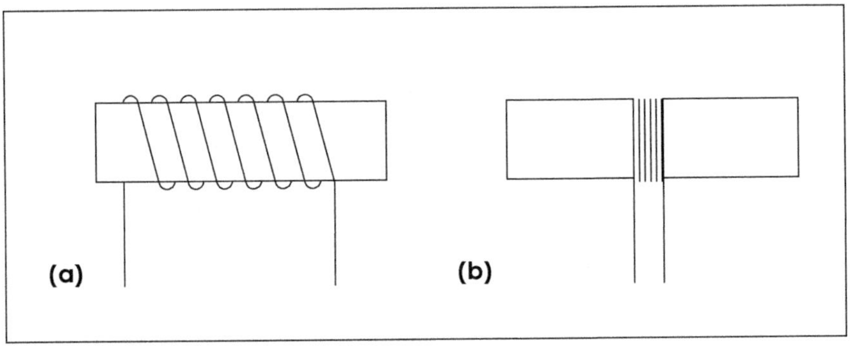

Figure 4. Encircling coil: (a) wide; (b) narrow.

For ferrous materials, a magnetizing coil holder is used (Figure 5). The magnetizing coil holder has steel guides at both the inlet and outlet sides that concentrate the magnetic field and provide magnetic saturation that eliminates the negative effects of permeability affecting the detection of discontinuities in the material. DC power is provided to the saturation coil while AC power is supplied to the eddy current coils.

Figure 5. Combination eddy current and ferrous inclusion test coils.

Figure 6 shows a magnetizing coil holder using permanent magnets. If more magnetization is needed, such as in the case of larger diameter test objects or heavier wall tubing, electromagnets may be used. With hotter material or continuous running tube inspection systems, an electromagnet with a water-cooling jacket is used to protect the coil windings from the excessive temperatures of the process. For ferromagnetic tubing inspection, demagnetization is often required and is accomplished through the use of an inline AC coil platform placed after the encircling coil testing. This demagnetizing coil can be adjusted as required to lower the residual magnetism caused during the magnetic saturation needed for testing.

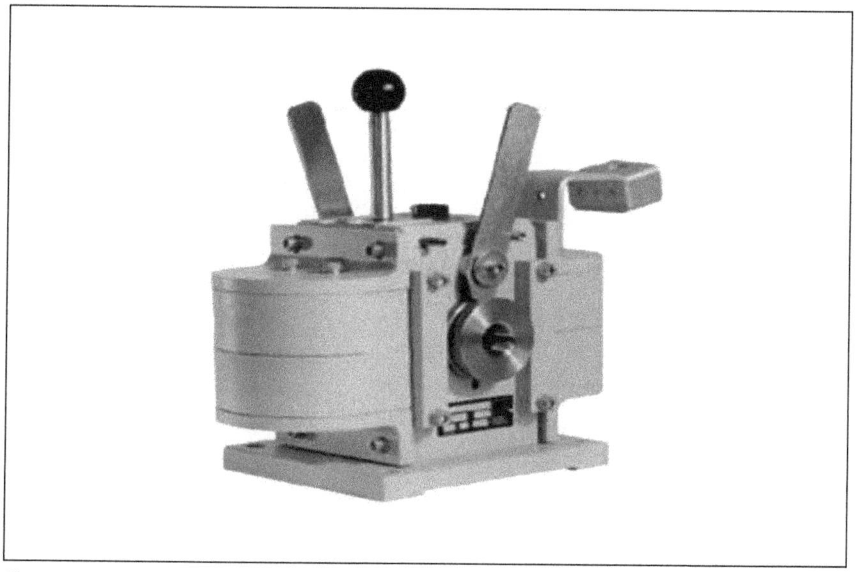

Figure 6. Magnetizing coil holder with permanent magnets.

Encircling Coil Theory

In eddy current encircling coil inspection, an alternating electrical current is provided at a fixed frequency into a test coil's primary field windings (excitation coil). This alternating current produces an alternating magnetic field. As the rod or wire passes through this magnetic field, eddy currents are induced into the test object around its circumference. These eddy currents generate a secondary magnetic field in opposition to the coil's primary magnetic field. The secondary magnetic field from the eddy currents induces a current into the coil's secondary measure windings (receiver coils). The electronics of the eddy current instrument interrogates these returned signals and compares them to defined threshold limits. If these signals exceed the predefined threshold limits, they are counted as discontinuities. The density and quantity of discontinuities are used to determine the rod or wire quality. Figure 7 shows the field windings (excitation coil) and the measure windings (receiver coils). The orientation of the magnetic field from the field windings is also shown.

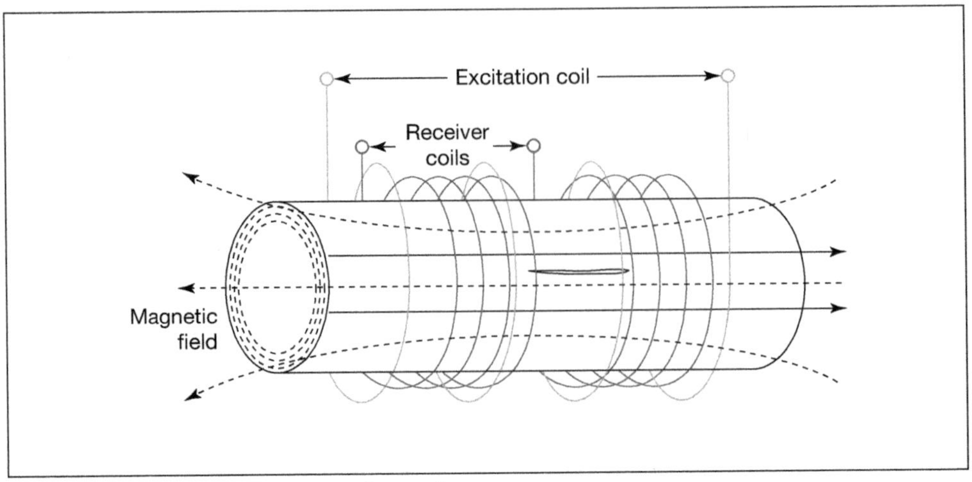

Figure 7. Eddy current coil configuration.

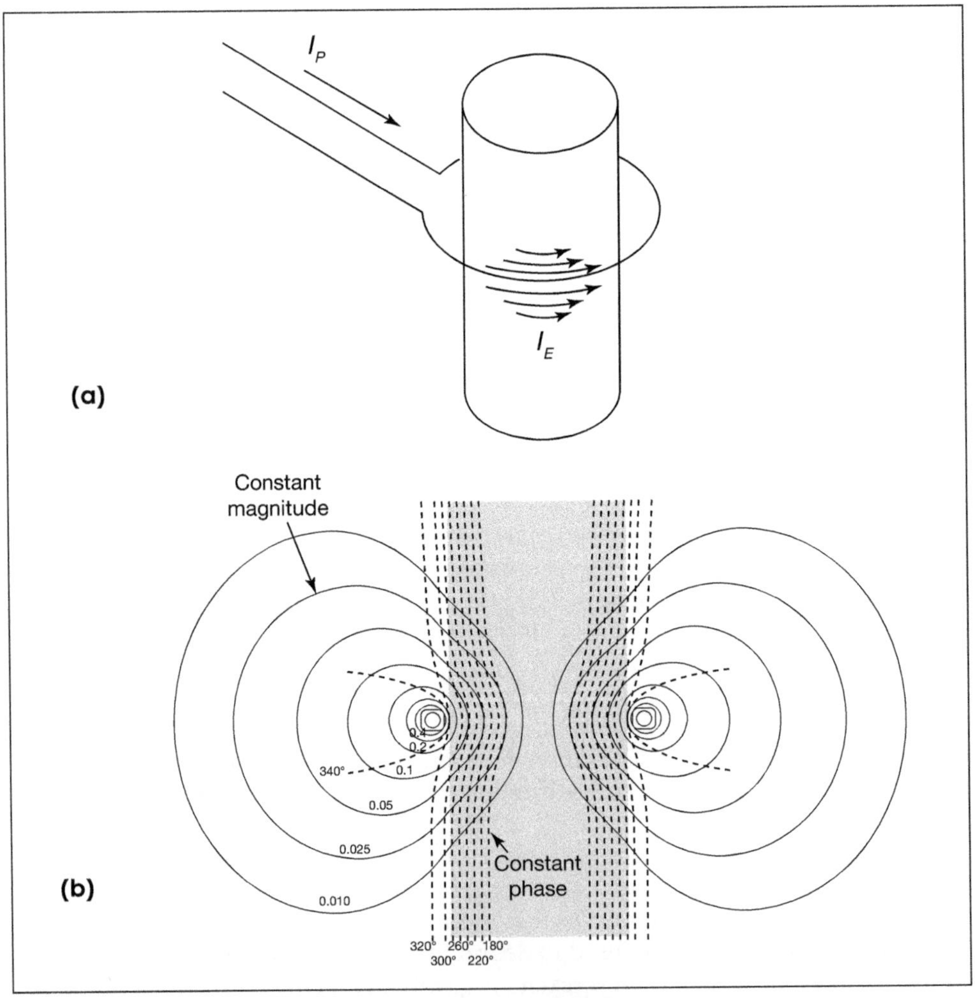

Figure 8. Magnetic field extension: (a) into a nonferromagnetic rod; (b) the opposing directional flow of the primary and secondary current.

Figure 8 shows the eddy current inspection region produced in a test rod by an encircling coil. Since the primary magnetic field intensity in the coil is normally considered to be constant across the diameter of the coil, it might seem reasonable to expect that the density of the eddy currents induced would also be constant across the rod. This is not true.

The eddy current's strength is greater near the surface and decreases toward the center of the rod. If the diameter of the rod is large enough, there is no current at the center at all. The phenomenon is known as the center effect and is the result of the opposition of the eddy current field. The eddy current magnetic field at the surface opposes the primary field and reduces it. This reduced field is again reduced by eddy currents just below the surface. Ultimately, the strength of the primary field is reduced to zero, and no additional eddy currents are induced. Note that the eddy current (I_E) in Figure 8a flows in the same plane as the currents flowing in the coil, that is, around the circumference of the rod but in the opposite direction. Thus, the encircling coil is especially adapted to locating discontinuities that are parallel or longitudinal to the length of the rod. With the current flow in the same plane as the coil is wound it is not possible to detect circumferentially oriented discontinuities with encircling coils. That discontinuity orientation would not disrupt the normal eddy current flow pattern.

Reference Standards

For tubing, saw cuts, drill holes, or electrical discharge machining (EDM), notches are used as setup standards (Figure 9). This initial setup and the baseline signal amplitude are consequently adjusted based on customer specific requirements and work instructions. The test object quality is then determined on the basis of how these signals compare to this established baseline signal response. An initial setup during mill operation may establish the instrument screen display amplitude at 15 to 20% for an assumed good test object. The baseline eddy current signal is adjusted with the sensitivity (gain) control.

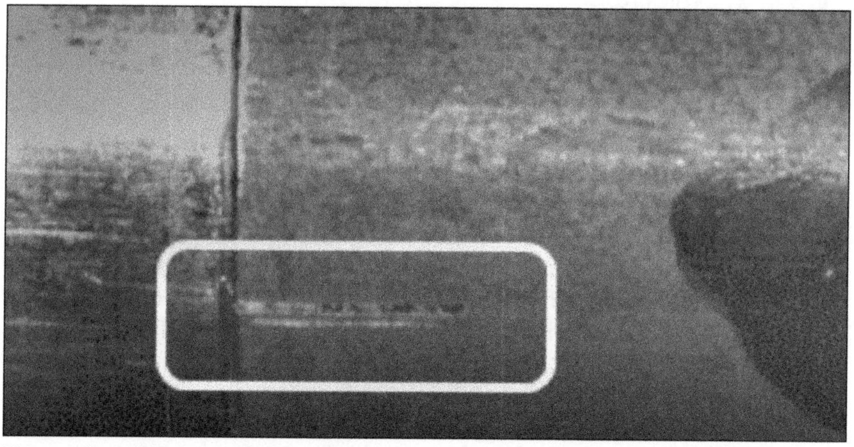

Figure 9. Artificial notch in steel tube used to optimize instrument setup. Dimensions of the notch are 0.01 in. (0.254 mm) deep by 0.1 in. (2.54 mm) long by 0.004 in. (0.102 mm) wide.

As material passes through the coil, the signal amplitudes are compared to this initial setup. If the test object quality is good, then the signals obtained from the test object are fairly uniform and consistent. Disruptions in the surface quality of the rod or wire create alterations in the returned signals from the coil's measure windings. Disruptions to this normal condition can be interpreted, depending on their severity and pattern, either as discontinuities or as deteriorating surface quality. (See Figure 10.)

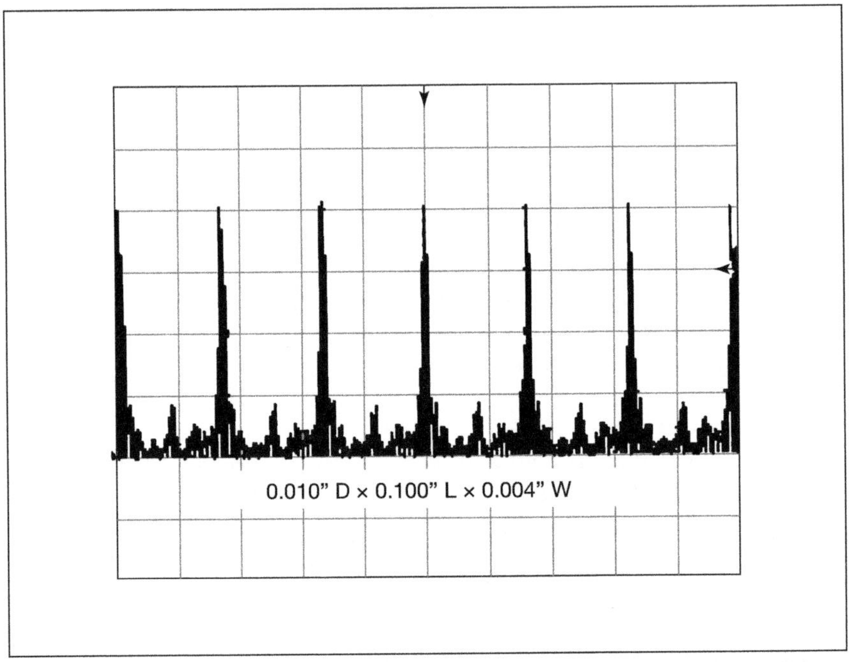

Figure 10. Digital display of the signal for the notch used in Figure 9 to optimize instrument setup. The printout shows that the reference object produced other smaller signals caused by other surface anomalies as the part was rotated.

The measure windings in the test coil are wound in a differential arrangement. Because one measure winding is continuously compared to the other measure winding, the coil is extremely sensitive to small surface cracks and quickly occurring discontinuities in the test object surface. Cracks and other sharply generated discontinuities cause an immediate and abrupt disruption to the flow of eddy currents in the rod or wire. Therefore, an immediate and abrupt increase to the eddy current signal amplitude will be displayed. Medium or large discontinuities in the material will trip the pre-established thresholds in the instrument's evaluation electronics.

Differential measure windings will, however, tend to balance on continuous longitudinal discontinuities such as drawing process scratches. These scratches show up as a cluster of higher density and lower amplitude signals. If the surface condition degrades gradually, the signal will also degrade gradually at an amplitude higher than the normal surface noise pattern. Discontinuity density analysis and counting methods may be used to accomplish detection of these conditions.

Test frequency selection depends on whether the depth of penetration or the discontinuity size is more critical. Typical test frequencies range from 10 kHz to 1 MHz; however, the actual application, material conductivity, and discontinuity specifications determine what is actually used. Higher test frequencies will detect smaller discontinuities but with less field penetration. Lower frequencies provide better field penetration but sacrifice some smaller discontinuity detectability. Both 30 and 100 kHz are successfully implemented in nonferrous rod and wire test lines.

Applications

Eddy current encircling coil inspection is performed without disruption to the manufacturing process. For this reason, encircling coil inspection is a very sensitive technique for detecting cracks, monitoring the general surface condition, and detecting discontinuities occurring on or near the surface of the material.

The encircling coil technique performs a continuous inspection of bar, wire, rod, and tubing during the actual manufacturing process. The test object passes through an encircling test coil and holder apparatus without contact. Since the rod and wire do not come into physical contact with the test coil, the inspection does not interrupt or slow down production. Since the encircling coil completely surrounds the test object, the entire outer circumference of the rod or wire is inspected. The advanced filtering capabilities available for the inspection do not impose any restrictions on the speed of the manufacturing process. Testing can take place at line speeds up to 20 000 ft/min (100 m/s).

Encircling coils are primarily used to test tubular and bar-shaped products. The tube or bar is fed through the coil, referred to as a feed-through coil, at relatively high speed. The cross section of the test object within the test coil is simultaneously interrogated. Since the eddy current flow is around the rod, circumferential discontinuities cannot be detected with this application.

Electromagnetic testing is a very effective tool for monitoring the integrity of the manufacturing process. Encircling coils used in the manufacturing process provide a mechanism to grade test objects, improve preventive maintenance, monitor process trends, and improve product quality. Electromagnetic testing can also provide a quality grading process as well as a statistically based production technique. The effective use of the information provided by the encircling eddy current technique can be used as part of ongoing process improvement. The process detects incremental changes in material conditions and provides the manufacturer with the required information to make informed decisions and take the appropriate corrective action.

Factors Affecting Choice of Sensing Elements

Several critical factors and specifications must be carefully considered when selecting eddy current sensing elements.

Frequency
The frequency determines eddy current depth of penetration and the amplitude and phase of a discontinuity response.

Excitation
The amplitude of the excitation signal determines the amplitude of the response. It should be well controlled, and its frequency response should be specified.

Gain Linearity
The amplitude and phase characteristics of gain stages must be qualified for adherence to a standard or specification appropriate to the application.

Horizontal and Vertical Deviation
Gain of the in-phase (X) and quadrature (Y) components of a signal must be controlled to prevent unwanted distortion of a signal. In a typical heat exchanger test, a nonlinear gain choice would cause misrepresentation of the data leading to errors in the phase analysis approach. However, there are cases where using nonlinear gains can be an advantage. For example, in the rotating probe test of a rivet hole for surface-breaking cracks, this technique can be used to minimize liftoff noise and accentuate a crack signal. The horizontal (noise) signal component is effectively reduced while the vertical (discontinuity) signal component is exaggerated.

Quadrature Accuracy
The phase of reference signals must be well controlled and the response of the display must be designed to ensure that the in-phase and quadrature components are truly at 90° electrically and they are displayed orthogonally.

Digitization Rate
The digitization rate for a digital system is the number of samples per unit of probe travel. This rate is critical for determining the response to a discontinuity. The digitization rate must be determined from the application criteria.

Sample Rate
The sample rate for a digital system is the number of interrogations per unit time, often given as samples per second. To calculate the required sample rate for a test, the probe speed is multiplied by the desired digitization rate.

Bandwidth
Bandwidth governs the response of the system as a function of frequency and is measured in hertz. A variable frequency, external modulator is often used to measure and characterize bandwidth through all stages of an instrument.

Alternating Current Field Measurement Probe Designs

Figure 11 shows components arranged in a typical alternating current field measurement test. The exact parameters used in a probe vary according to the application. The larger dimensions are used where possible because they give the most uniform field and allow the two sensors to be wound concentrically, which gives clear symmetric loops in the data presentation. In probes designed for tight access applications or for higher sensitivity, the smaller dimensions are used.

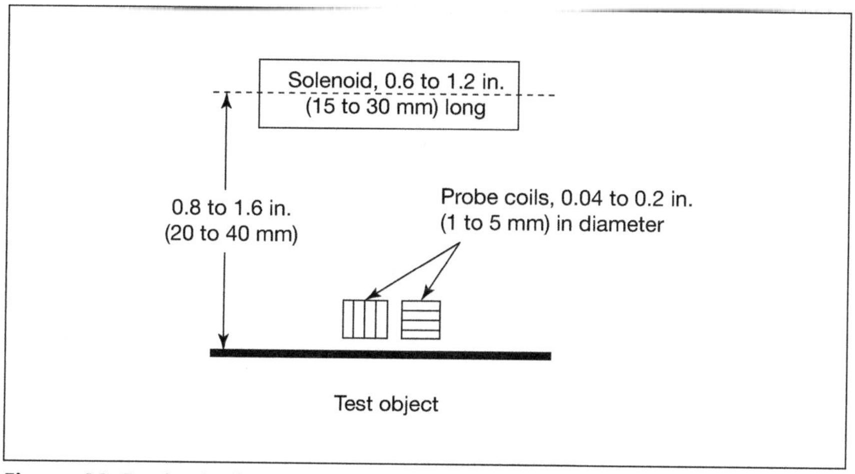

Figure 11. Typical alternating current field measurement probe layout.

Uniform Field

A uniform field has advantages and disadvantages compared with conventional eddy currents. These advantages and disadvantages are discussed as follows.

Advantages

Testing through Coatings

The primary advantage of using a uniform field is that the intensity of the input field decays gradually with distance from the inducing coil; the intensity of the field perturbed by a discontinuity also decays gradually with distance above the surface. The intensity of a uniform field indication does not drop off very rapidly with probe liftoff, so alternating current field measurement can be used to test through thick nonconductive coatings. The technique can be used on painted or rusty surfaces or on structures covered with protective or fire-resistant coatings one- or two-tenths of an inch (several millimeters) thick.

Depth Information

The second advantage is that the larger inducing coil forces currents to flow farther down the face of a deep crack. When a conventional eddy current probe lies over a deep crack, the current splits into two separate circles, one on each side of the crack, confined to approximately 0.12 in. (3 mm) of the crack face. Because essentially no current then flows to the bottom of the crack, no information can be obtained about where the bottom is, so the depth of the crack cannot be measured. The same feature occurs with an alternating current field measurement probe. However, because the depth of penetration down the crack face is related to the size of the magnetic-field-inducing coil, an alternating current field measurement probe can measure more deeply, typically 0.6 to 1.2 in. (15 to 30 mm), depending on probe type.

Material Boundaries

A third advantage of a uniform field arises when testing at a weld or other boundary between two metals of different permeability or conductivity. In this case, if the probe is scanning for discontinuities parallel to the boundary, no probe motion is required across the boundary and no signals are caused by the change in material property. Also, the currents are flowing perpendicularly across the boundary, so the effect of this material change is reduced even when scanning up to it.

Disadvantages

Reduced Sensitivity

The main disadvantage of using a uniform field is that sensitivity is reduced. This reduction is of little consequence on welded or rough surfaces, where sensitivity would be reduced anyway. On smooth, clean surfaces, however, alternating current field measurement is less sensitive to short or shallow discontinuities than conventional eddy current techniques. The smallest detectable discontinuity on a good surface with alternating current field measurement is approximately 0.08 in. (2 mm) long or 0.01 in. (0.25 mm) deep.

Geometry Changes

A second disadvantage of a uniform field is that, because the currents spread out farther, signals are obtained from local geometry changes, such as plate edges and corners. Although these signals do not usually have the same form as a signal from a discontinuity, they can confuse the operator. If many similar geometries are being tested, the operator can learn what signals are caused by the geometry alone and then ignore these. Alternatively, scans from discontinuity-free sites with the same geometry can be stored and displayed for comparison or probes with differential sensors can be used to eliminate the large-scale signals.

Discontinuity Orientation

A third disadvantage is that the signals obtained from a discontinuity depend on the orientation of the discontinuity. The uniform field theoretical model suggests that no signal is produced when a probe scans across a transverse discontinuity because the current flow is then parallel to the discontinuity and is not perturbed. In practice, there is a signal produced in this situation (caused by magnetic flux lines "jumping" the discontinuity), but these do not conform to the signal expected from a discontinuity. The operator is trained to look for the signals caused by a transverse discontinuity in order to detect them. Additional scans must be made along the line of the discontinuity to size it.

Remote Field Testing Sensing Elements

Probe Configuration

In the basic remote field testing probe (Figure 12), there is one exciter coil and one detector coil. The dimensions of the coils vary from manufacturer to manufacturer. Both coils are wound circumferentially with respect to the tested tube and are separated by a distance greater than twice the tube diameter. This axial distance is characteristic of remote field testing. If the exciter and detector were to be placed close together, the detector would measure only the field generated by the exciter in its vicinity. In that case, the setup would basically be a standard eddy current test setup in send-and-receive mode.

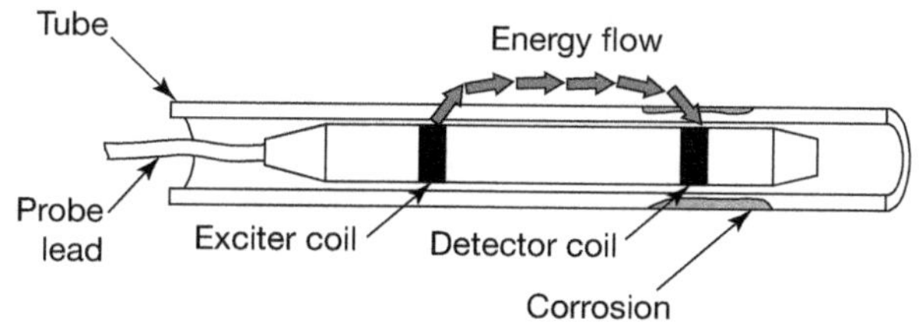

Figure 12. Simple probe for remote field testing.

To observe remote field testing's unique through-wall transmission effect, the inspector needs to move the detector away from the exciter. The actual separation depends on the application and the probe manufacturer but will always be a minimum of two pipe diameters. It is this separation that gives remote field testing its name: the detector measures the electromagnetic field remote from the exciter. Although the fields have become very small at this distance from the exciter, they contain information on the full thickness of the pipe wall.

Remote field testing probes often contain two or more receiver coils. The coils are connected to a remote field testing instrument by coaxial cable, where the outer conductor is used to shield the inner conductors from ambient noise. The coaxial cable is usually housed in a stiff plastic tube that lets the probe be pushed into a heat exchanger tube or pipe for distances up to 100 ft (30 m).

Features of Remote Field Testing

Remote field testing operates at relatively low frequencies. Typical frequencies are in the range of 40 to 500 Hz. The test speed for carbon steel tubes is approximately 30 ft/min (150 mm/s). A two-person crew can examine from 200 to 500 tubes measuring 30 ft (9 m) long in an 8 h shift, depending on access, setup time, number of discontinuities encountered, and other factors. Remote field testing is a noncontact technique, so the probes have minimal friction with the pipe wall and require no couplant.

Sensitivity

The accuracy for remote field testing in the straight part of the tubes is approximately 10% of wall thickness for general wall loss. The accuracy is generally less (20% of wall thickness) for highly localized discontinuities and in bends or near external conducting objects. This is due to the changes in magnetic properties of the tube in the bend area and because of shielding effects of external objects.

Remote field testing is equally sensitive to inside and outside surface discontinuities but usually cannot discriminate between them without the help of near field coils. Remote field testing is relatively insensitive to scale and magnetic debris.

Effects of Probe Speed

The ability of remote field testing to detect and quantify discontinuities relies on the quality of the signal received. The probe must be in the vicinity of the discontinuity for at least one excitation cycle in order to detect it. Ideally, the probe senses a discontinuity for several cycles so that any noise signals can be averaged out of the data.

If a test frequency of 100 Hz is being used, there are 100 opportunities per second to measure the signal. If the smallest discontinuity that is required to be detected has a length of 0.12 in. (3 mm), then the fastest speed possible is 12 in./s (300 mm/s). To build in a safety margin, the test speed should be no faster than half this speed. The probe pull speed should be slow enough so that the digital sample rate allows the field profile near the probe to be accurately recorded. It is equally important that the speed of testing be as constant as possible. Sudden changes in speed can result in anomalous signals.

Special Probes

To improve sensitivity of the remote field technique to discontinuities close to a tube support plate or tube sheet, probes can contain detector coils on each side of an exciter coil. Alternatively, two exciters can be used with one or two detectors placed between them. The disadvantage of these probes is their increased length and more complex data.

Pipes with diameters larger than 1 in. (25 mm) often require more than one detector coil to improve detection of local discontinuities. Array coils are used to segment the detector section to improve small-volume discontinuity sizing. The probe in Figure 13 is being used to examine a 3 in. (75 mm) diameter pipeline used to transport gas from the wellhead.

Array probes require special software to display and analyze the large quantities of multichannel data. It is possible for software to analyze remote field test data automatically. In all cases of automatic data analysis, the success of the analysis depends on the criteria that the operator uses to train the software to recognize signals of interest. It is also very important to have a skilled technician review a percentage of the calls made by automatic analysis software to verify accuracy. Only one selected channel at a time can be displayed on a voltage plane display.

Figure 13. Array probe in use on 3 in. (75 mm) pipeline.

7

Magnetic Flux Leakage Sensing Elements

Inductive Coil Sensors

Pickup Coils

In order for an inductive coil to detect perturbations in a static magnetic field, there must be relative motion between the coil and the field such that the flux within the coil changes. One of the simplest and most popular means for detecting leakage fields is to use a pickup coil. Ferrites are useful in pickup coils because they do not only provide support for the wire turns but they also concentrate the flux density through the coil windings by a value equal to the effective permeability of the ferrite.

For small pieces of ferrite (Figure 1) where the dimensional ratio is small, the effective permeability of the ferrite may vary from the low teens to the thousands. The advantage of using ferrite in this application is that ferrites have very low electrical conductivities, minimizing detrimental eddy current effects.

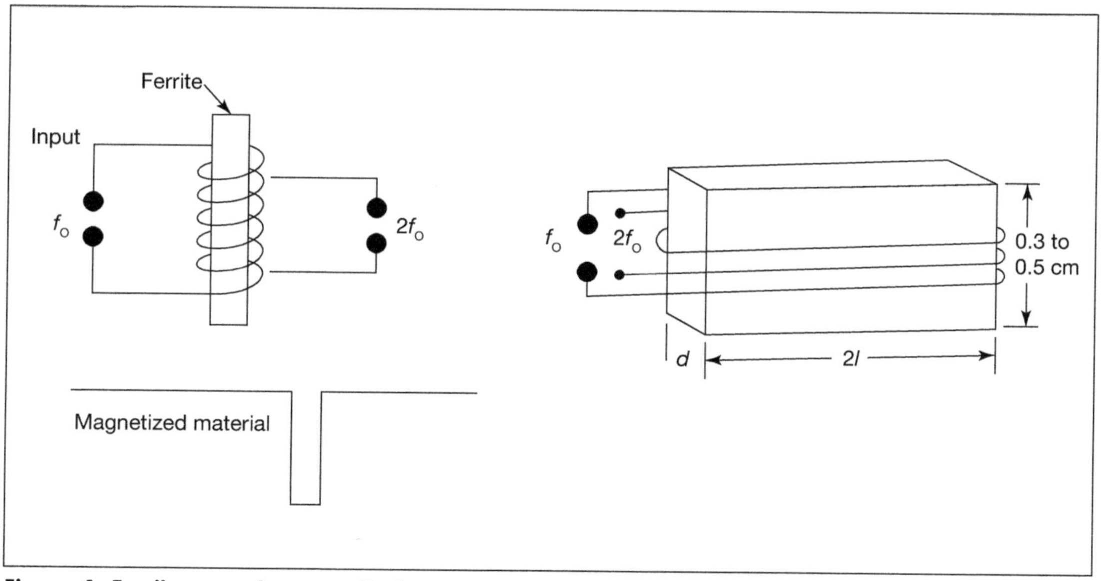

Figure 1. Ferrite-cored magnetic flux leakage detector coil systems.

It is important to note in the inductive sensor approach that the flux density must be changing through the coil in order to produce a signal.

It is essential that pickup coils be used to generate voltages and not currents. Once a current is allowed to flow in a coil, it creates its own magnetic field—one that can interfere with the field under investigation. The output of such coils is therefore generally fed to a high-resistance operational amplifier.

Flux Leakage Signals

Leakage fields tend to be local and concentrated near discontinuities. The leakage field can be divided into three orthogonal components: normal (vertical), tangential (horizontal), and axial directions. Probes are usually designed or oriented to measure one of these components. The voltage induced in the coil is given by the rate of change of flux. The output of the pickup coil is proportional to the spatial gradient of the flux along the direction of the coil movement as well as the velocity of the coil. Figure 2 shows the output of a pickup coil and the signal obtained after integrating the output. The coil is used to measure, in units of gauss (tesla), the magnetic flux density (*B*) leaking from a rectangular slot.

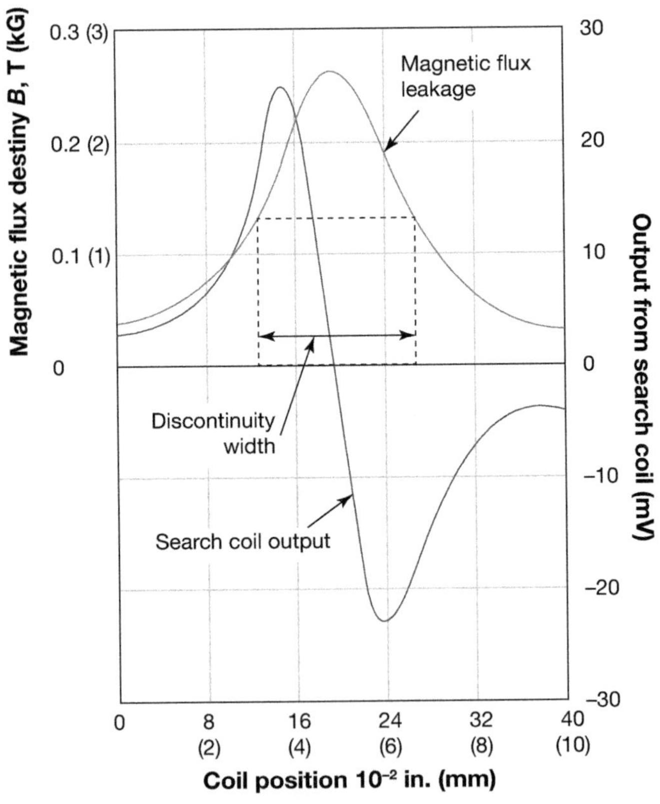

Figure 2. Output signal of a magnetic flux leakage pickup coil.

Hall Effect Sensors

Hall elements are crystals of semiconductor material. When a current is passed through them and they are placed in a magnetic field, a voltage develops across two of the faces of the crystal. The voltage is proportional to the strength of the magnetic field.

Bulk hall elements are generally bismuth-doped semiconductors such as indium antimonide (InSb). These are produced by solid-state crystal growth technology, cut into small rectangular blocks, and attached with current and voltage leads before being encapsulated. Typical sizes are as small as 0.03 in. (0.08 cm) long by 0.015 in. (0.04 cm) wide by 0.02 in. (0.05 cm) thick.

Vapor-deposited hall elements have been used in the testing of ball bearings using the magnetic flux leakage method. In this application, bismuth is evaporated onto an alumina substrate. A newer development is to combine the hall sensor, its power supply, and an amplifier on one chip. Figure 3 and Table 1 show configurations of typical hall sensors and their specifications.

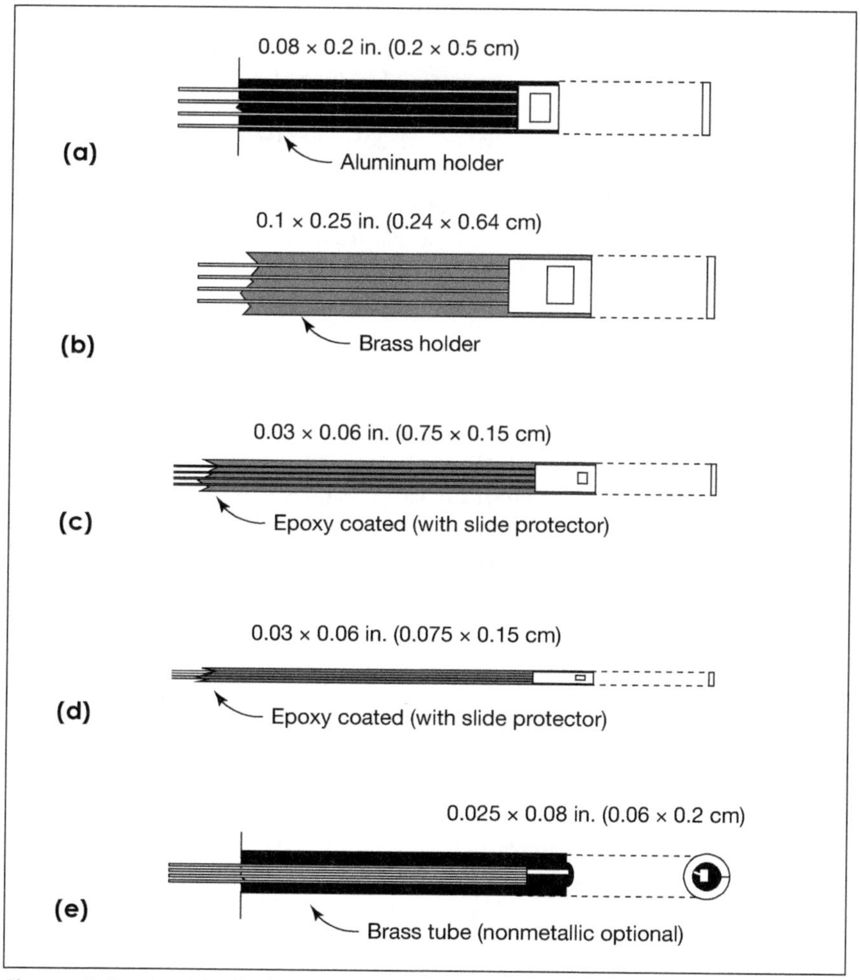

Figure 3. Typical hall element probes: (a) flat; (b) high linearity; (c) miniature; (d) subminiature; and (e) axial.

Table 1. Specifications of typical hall effect probes

Probe type	Hall output voltage (mV)	Nominal current control (mA)	Temperature coefficient (°C)	Operating temperature (°C)
Flat or transverse	340	200	–0.1	–65 to 85
High linearity	350	350	–0.1	–65 to 85
Miniature	200	25	–0.25	–65 to 85
Subminiature	200	25	–0.25	–65 to 85
Axial	100	100	–0.1	–65 to 85

Since hall effect sensors do not depend on motion for their sensitivity, they can be scanned at any rate that is mechanically convenient, which may present an advantage over inductive coil sensors depending on the application. However, hall effect sensors are difficult to fabricate, are somewhat delicate, and require more complex auxiliary electronic apparatus than inductive coil sensors.

Flux Gate Magnetometer

The flux gate magnetometer, also referred to as a ferroprobe or foerster probe, is a device that measures magnetic fields by utilizing the nonlinear magnetic characteristic of ferromagnetic core materials as its sensing element. A drive coil and sense coil are wound onto a saturated core. The core characteristics and drive current are such that the magnetization changes induced by the leakage field affect the filter harmonic output of the sense coil. In operation, the ferromagnetic core of the sensor is driven cyclically to saturation by means of a periodic current of suitable wave shape in the drive coil windings.

In the absence of a signal field, which is usually direct current or very low-frequency alternating current, and is generally a very small fraction of the peak value of the driving field, the voltage induced in the sense winding is symmetrical. In other words, it contains only odd harmonics of the fundamental of the drive current. In the presence of a signal, the sense winding voltage becomes asymmetrical. This asymmetry is sensitively related to the signal field and can be detected by various techniques.

Magnetodiode

The magnetodiode is a solid-state device, the resistance of which changes with magnetic field intensity. It consists of *p* zones and *n* zones of a semiconductor, separated by a region of material that has been modified to create a recombination zone, as illustrated in Figure 4.

Active areas typically measure 0.12 × 0.024 × 0.016 in. (0.3 × 0.06 × 0.04 cm), and output signals are generally larger than for hall elements, although the response to field intensity is not so linear for higher fields, as shown in Figure 5.

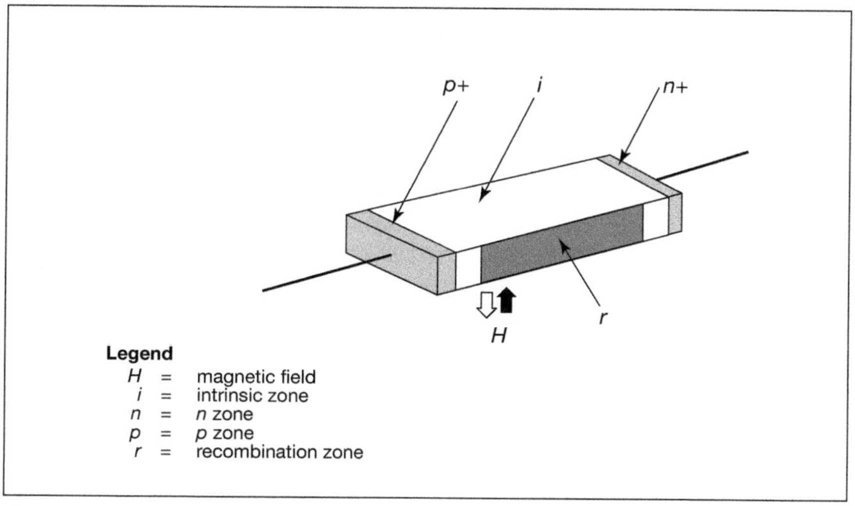

Figure 4. Schematic of a magnetodiode.

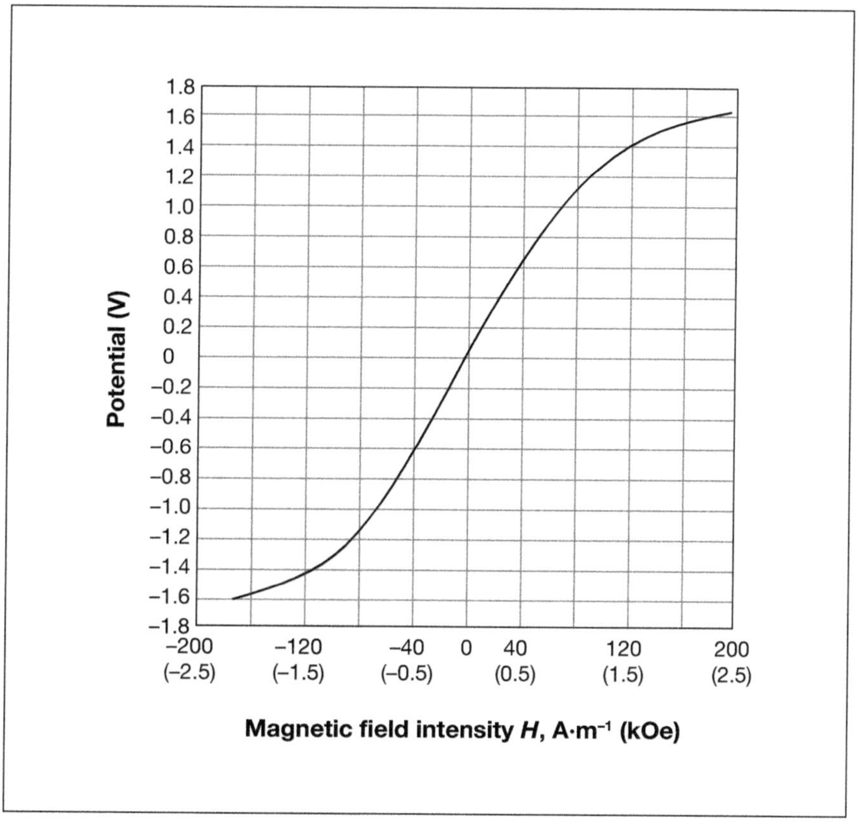

Figure 5. Response of magnetodiode; linear up to approximately 40 kA/m at ambient temperature of 77 °F (25 °C) and potential of 1.6 V.

Applications of Magnetodiodes

Figure 6 shows the use of magnetodiodes for detecting magnetic flux leakage from discontinuities in tubes. The magnetic flux leakage is excited by alternating current electromagnets arranged to detect either internal or external surface-breaking discontinuities. The system illustrates the general principles of magnetic flux leakage testing. Sensors are connected differently to eliminate signals from the applied field and from relatively long-range variations in surface field strength. This system and similar magnetic flux leakage systems are used to rapidly evaluate the surface condition with a depth of only 0.004 in. (0.01 cm).

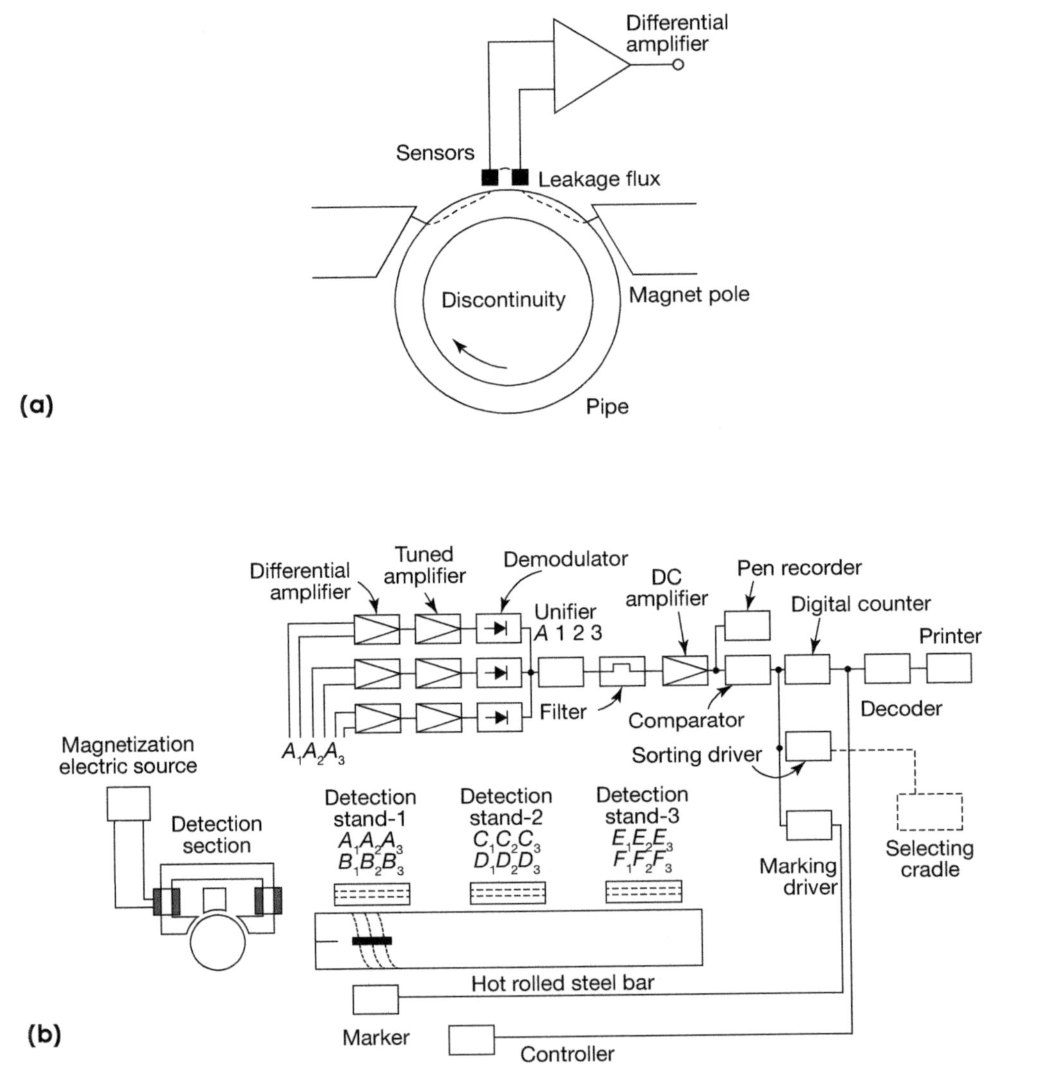

Figure 6. A magnetodiode testing system for tubes: (a) alternating current magnetizing technique; (b) electrical block diagram.

Other Magnetic Flux Leakage Field Detection Methods

Magnetic Tape System

For the testing of flat plates and billets, it is possible to scan the surface with wide strips of magnetic recording tape. Discontinuity signals are taken from the tape by an array of tape recorder heads. Scale, dirt, or oil on the test surface can contaminate the tape. Surface roughness can tear the tape.

Magnetic Resonance Sensors

Nuclear magnetic resonance magnetometers are based on the fact that the characteristic atomic frequencies also depend on the strength of magnetic field. In operation, when an atomic nucleus is placed in a constant magnetic field and is subjected to a high-frequency alternating magnetic field, resonance absorption of energy from the alternating current field takes place.

The absorption always takes place at a fixed ratio of constant field strength to the alternating field frequency. Hence, if the frequency at which nuclear magnetic resonance takes place is measured, the constant field strength can be determined. Resonance magnetometers include a large family of devices that can measure magnetic fields of any strength and are subject to a wide variety of applications in the engineering field.

8

Conductivity, Coil Impedance, and Coupling

There are three fundamental properties of material that affect the eddy currents induced in the test part:

- The conductivity of the test object.
- The permeability of the test object.
- The dimensions of the test object.

Conductivity Factors

Conductivity of a material is defined as the ability of the material to carry electrical current, that is, the number of amperes of current that will flow through a given size (cross-sectional area) of the material when a given voltage is applied to the material. This definition is too cumbersome to use with ease in electromagnetic testing. A simplified approach to classifying materials is based on the International Annealed Copper Standard (IACS) system. The symbol that is commonly used for conductivity is σ (sigma) and the units can be expressed as a percentage of IACS.

In the IACS system, the conductivity of unalloyed (pure) annealed copper was selected as the standard. The conductivities of all other materials are expressed as a percentage of this standard.

Unalloyed annealed copper is assigned a rating of 100% IACS, and a material that conducts electrical current only half as well is rated at 50% IACS. For example, a wire made of aluminum can carry only 61% of the current that can be carried by the same size wire made of pure alloyed copper at a given voltage.

Table 1 lists the IACS ratings of several materials. The table illustrates that materials do have different abilities to conduct electrical current.

Good conductors include copper and silver, poor conductors include titanium and stainless steel, and nonconductors include wood and glass. Differences in the conductivity of different materials are detectable by electromagnetic testing due to the effect that the conductivity of the material has on the magnetic field of the coil.

Resistivity is defined as the ability of material to resist the flow of current. The symbol for resistivity is ρ (rho) and the units are sometimes expressed in micro-ohm centimeters (μΩ·cm). Conductivity and resistivity work in opposition to each other. Conversions between conductivity and resistivity can be made using simple division. The original copper bar used to establish the standard had a direct current

Table 1. Relative conductivity of various metals and alloys

Metal or alloy	Conductivity (%IACS)
Silver	105
Copper, annealed	100
Gold	70
Aluminum	61
Aluminum alloys: 6061-T6 7075-T6 2024-T4	 42 32 30
Magnesium	27
70-30 brass	28
Phosphor bronzes	11
Monel	3.6
Zirconium	3.4
Zircaloy-2	2.4
Titanium	3.1
Ti-6Al-4V alloy	1.0
304 stainless steel	2.5
Inconel 600	1.7
Hastelloy X	1.5
Waspaloy	1.4

resistance of 0.017241 ohms. A conversion factor of 172.41 is used to adjust values of either conductivity or resistivity into the other form.

To convert to either unit, use Equation 1.

(Eq. 1)

$$\%\text{IACS} = \frac{172.41}{\mu\Omega \cdot \text{cm}}$$

As resistivity increases, conductivity decreases, and vice versa.

Problem 1

Convert resistivity of 5.5 micro-ohm centimeters to conductivity in %IACS.

Solution

$$\%\text{IACS} = \frac{172.41}{5.5\ \mu\Omega \cdot \text{cm}}$$
$$= 31.3$$

Table 2 lists the electrical conductivity and resistivity of some common metals and alloys.

Table 2. Electrical resistivity and conductivity of selected metals and alloys

Metal	Conductivity		Resistivity	
	MS·m^{-1}	%IACS	Ω·m	μΩ·cm
Aluminum, pure	35.38	(61)	2.83×10^{-8}	(2.83)
Aluminum (99.99%)	37.67	(64.94)	2.65×10^{-8}	(2.65)
Antimony	2.55	(4.4)	3.92×10^{-7}	(39.18)
Bronze, commercial annealed	25.52	(44)	3.92×10^{-8}	(3.92)
Cadmium	14.62	(25.2)	6.84×10^{-8}	(6.84)
Calcium	28.25	(48.7)	3.54×10^{-8}	(3.54)
Chromium	5.10	(8.8)	1.96×10^{-7}	(19.59)
Cobalt	16.01	(27.6)	6.25×10^{-8}	(6.25)
Copper	58	(100)	1.72×10^{-8}	(1.72)
Gold	40.60	(70)	2.46×10^{-8}	(2.46)
Iron, pure	10.44	(18)	9.58×10^{-8}	(9.58)
Iron ingot (99.9%)	9.05	(15.6)	1.11×10^{-7}	(11.05)
Magnesium, pure	22.39	(38.6)	4.47×10^{-8}	(4.47)
Molybdenum	19.14	(33)	5.22×10^{-8}	(5.22)
Nickel	14.62	(25.2)	6.84×10^{-8}	(6.84)
Selenium	8.35	(14.4)	1.20×10^{-7}	(11.97)
Silver, tin solder	9.63	(16.6)	1.04×10^{-7}	(10.39)
Steel, high alloy	1.68	(2.9)	5.94×10^{-7}	(59.45)
Tin, pure	8.70	(15)	1.15×10^{-7}	(11.49)
Tin foil	2.44	(4.2)	4.10×10^{-7}	(41.05)
Tungsten	18.21	(31.4)	5.49×10^{-8}	(5.49)
Zinc, commercial rolled	16.24	(28)	6.16×10^{-8}	(6.16)

Factors Affecting Conductivity

Even though the chemistry or the alloy content of a material may be generally known, there are multiple factors that can change the actual conductivity of a given piece of metal. Several of these factors are described as follows.

Alloy Composition

Alloys are combinations of metals. Each metal has a specific chemistry with a range of characteristics. When multiple base metals are combined and then processed (heat treated), the end product is an alloy with some but perhaps not all of the characteristics of the original base metals.

The new alloy can be chemically defined by its makeup. There are groups of materials that can be assumed to be similar, such as the copper alloys (90-10 and 70-30 Cu-Ni) or the 300-series stainless steels (304-SS, 316-SS, and 316L-SS). However, the chemistry alone will not give us the conductivity value of a given sample. Aluminum, for example, has a wide range of potential values. Depending on final alloying and heat-treat processes, a piece of aluminum could fall anywhere between 25 and 65% on the IACS scale.

Hardness

When a metal is subjected to heat treatment, it will become either harder or softer depending on the material and the processes involved. This change in hardness is brought about by the internal change in the crystalline structure. This change at the crystalline level will alter the conductivity and/or permeability of the material. These changes can be detected by electromagnetic testing. An improper heat treatment can be detected with this method.

Temperature and Residual Stresses

The ambient temperature and internal residual stresses of a test material also have an effect on the conductivity of the material. These changes can be detected by electromagnetic testing. An increase in the temperature of the material normally results in a decrease in the conductivity of the material. Residual stresses cause an unpredictable, but detectable, change in conductivity.

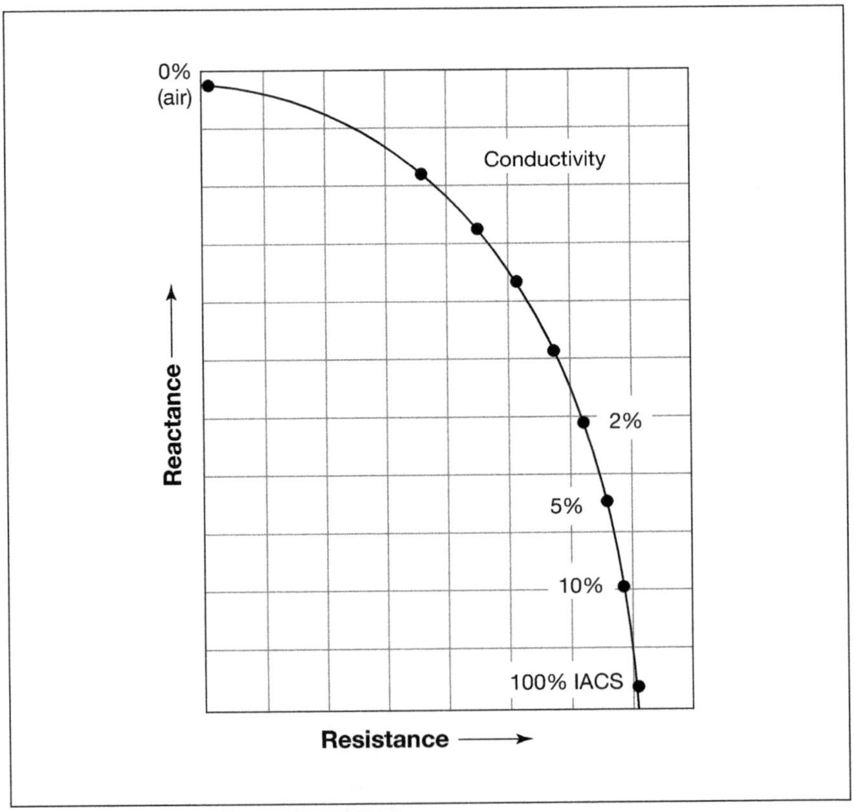

Figure 1. Measured conductivity locus.

Conductivity Coatings

The presence of a conductive coating on a conductive material changes the inherent conductivity of a base metal just as an alloy would. However, if the thickness of the cladding varies, the conductivity will vary. This change in thickness can be detected using eddy current testing methods.

As the test coil is influenced by different conductivities, its impedance varies inversely to conductivity. A higher conductivity causes the test coil to have a lower impedance value.

The coil's inductive reactance is represented by the Y axis, and coil resistance appears on the X axis. The 0% conductivity point, or air point, is when the coil's empty reactance (X_{L0}) is maximum. Figure 1 represents a measured conductivity locus. The measured conductivity or localized conductivity changes (for example, due to discontinuities) can be affected not only by the material factors previously discussed but also by the following test environment factors, namely, edge, skin, and end effects.

Edge Effects

Edge effect is the term used to discuss the impedance change of probe coils when a geometrical edge is approached. The coil senses less and less conductivity as it approaches the edge of the test object. Edge effects can be demonstrated by moving the eddy current probe toward the edge of a test sample and observing the change in instrument reading that results. Figure 2 shows an example of instrument change resulting from edge effect. Test samples must be large enough to prevent this edge interference, or probes must be shielded to collimate the field. In this particular case, an edge distance of 0.2 in. (0.5 cm) or more must be maintained to avoid errors. If test objects have a narrow width, a centering jig or holder should be used to maintain the probe on center. Edge distance curves are used to apply correction factors to conductivity readings on production parts. Even shielded probes have some edge effect.

Skin Effect

In many applications, electromagnetic tests are most sensitive to test object variables nearest to the test coil due to the skin effect. The skin effect is a result of mutual interactions of eddy currents, operating frequency, test object conductivity, and permeability. The skin effect (the concentration of eddy currents in the test object nearest the test coil) becomes more evident as test frequency, test object conductivity, and/or permeability are increased.

End Effect

End effect is a term most applicable to the testing of bar or tube products. The end effect follows the same logic as the skin effect. The end effect is the signal observed when the end of a product approaches the test coil. Response to the end effect can be reduced by coil shielding or reducing coil length in outside diameter encircling or inside diameter bobbin coils.

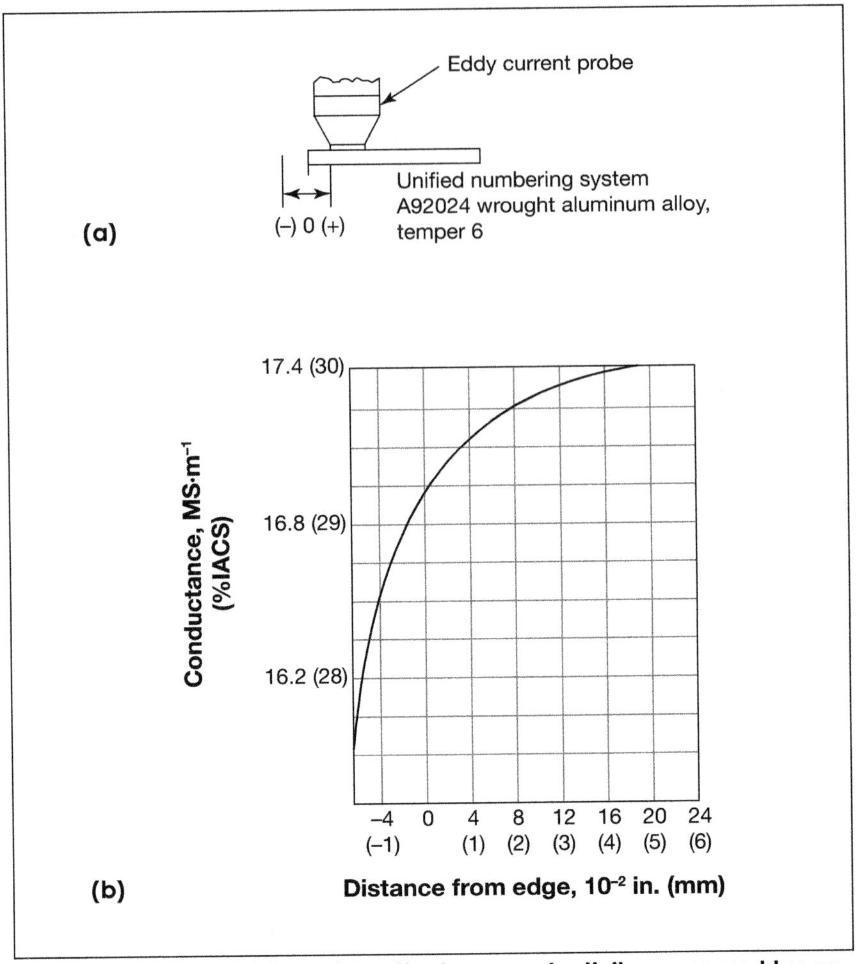

Figure 2. Influence of the edge effect on conductivity measured by an eddy current probe: (a) probe at edge; (b) signal.

Permeability Factors

Permeability originates from the word "permeate," meaning to spread through. Permeability is the ease with which a material can become magnetized.

Soft iron and iron with low carbon content are very easy to magnetize and are highly permeable. These ferromagnetic materials readily conduct the lines of force. Ferromagnetic materials that are hard to magnetize have low permeability. Hardened ferromagnetic steel with high carbon content is hard to magnetize and has low permeability.

Nevertheless, this type of steel will hold some of the magnetism after the magnetizing current is shut off. That is how a permanent magnet like the horseshoe magnet is made. The magnetism retained in a magnet is called residual magnetism.

When an energized test coil is placed on ferromagnetic materials, the field is greatly intensified by the magnetic properties of the material so that a large change in the impedance of the test coil occurs.

If the magnetic field strength varies even slightly, these small variations have a large effect on the impedance of the coil. These changes in the impedance of the coil are often so large that they mask all other changes, such as conductivity and dimensional changes. This effect may be overcome by magnetizing the material to saturation using a separate coil energized by a direct current source. Magnetic saturation eliminates or reduces any variations in the residual magnetic field caused by magnetic variables, and thus allows other variations to be measured. Electromagnetic tests on carbon steel materials where saturation is not used limits the depth of penetration to very small dimensions. During saturation, the slope of the *B* and *H* curve is almost horizontal, and the permeability relative to air is approximately 1.

Dimensional Factors

Dimensional factors of the material that are of concern to electromagnetic testing fall under two types:

- The dimension and shape of the test object.
- The presence of discontinuities in the test object.

Test Object Shape and Thickness

Eddy currents do not penetrate throughout thick material but tend to be concentrated near the surface. Thus, there is a finite, or limited, depth of penetration. For mathematical reasons, it has become useful to define the standard depth of penetration as the distance from the surface of the test object to the point where the current density is approximately 37% of the current density at the surface.

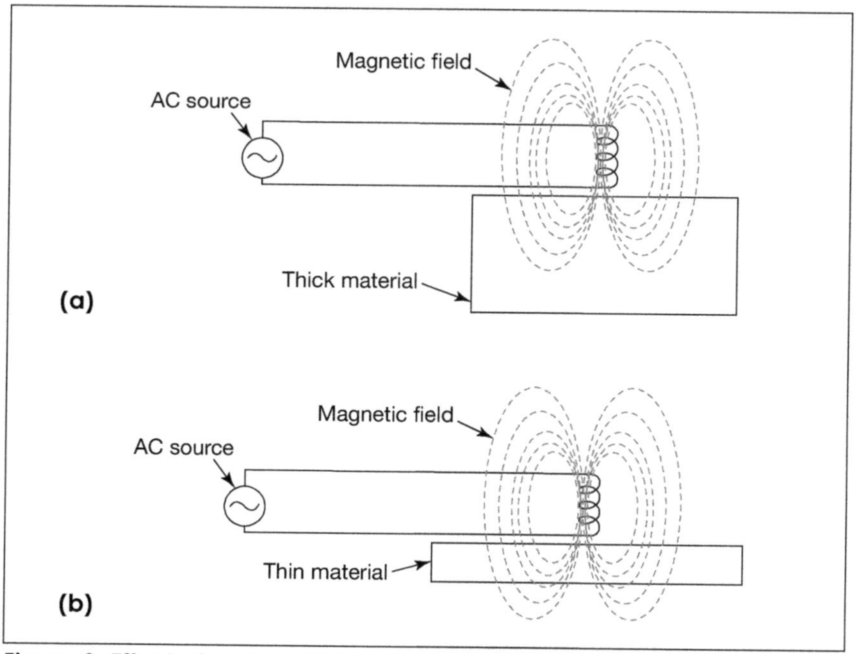

Figure 3. Effect of material thickness on eddy currents: (a) for thick material; (b) for thin material.

The depth of penetration of eddy currents in a nonmagnetic test object depends on the conductivity of the nonmagnetic material (the greater the conductivity, the less the penetration) and the frequency of the alternating current used to energize the test coil (the lower the frequency, the greater the penetration).

When the material is thin enough (as shown in Figure 3) so that all of the coil's magnetic field is not used in creating eddy currents, the strength of the eddy currents is reduced. This appears in the test circuit as an apparent difference in conductivity from that of the thicker piece of the material.

Discontinuities

A discontinuity is defined as any interruption in the normal physical structure or configuration of an object. The flow of eddy currents within a test object is affected by the presence of discontinuities, such as cracks, pits, vibrational damage, and corrosion. Discontinuities in a test object disturb the normal eddy current flow, and this results in a change in the coil impedance.

The magnitude of the indication caused by a discontinuity primarily depends on the amount of current disrupted by the discontinuity. In other words, the depth, width, and length of a discontinuity determine the change in the eddy current flow, as shown in Figure 4.

Discontinuities open to the surface are more easily detected than subsurface discontinuities. Discontinuities open to the surface can be detected with a wide range of frequencies; subsurface investigations require a more careful frequency selection.

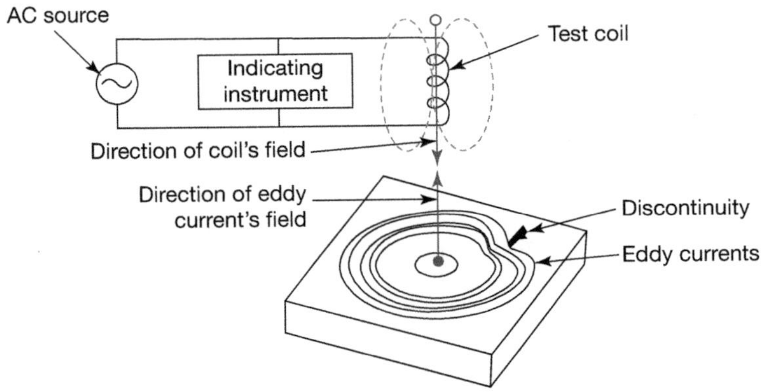

Figure 4. Distortion of eddy currents by a discontinuity.

Liftoff and Fill Factor

Coupling refers to the effectiveness of a coil in inducing eddy currents into a test object. Two electric circuits are regarded as coupled when having an impedance in common so that a current in one causes a voltage in the other. Two coils are said to be coupled when they share parts of their magnetic flux paths. With regard to a coil coming in contact with a test object, coupling is dependent on the amount of liftoff and on the fill factor.

Liftoff

Liftoff describes any space that occurs between the test object and the inspection coil. When a surface coil is energized and held in the air above a conductor, the impedance of the coil has a certain value. As the coil is moved closer to the conductor, the initial value will change when the field of the coil begins to intercept the conductor. Because the field of the coil is strongest close to the coil, the impedance value continues to change until the coil is directly on the conductor. Conversely, once the coil is on the conductor, any small variation in the separation of coil and conductor will change the impedance of the coil.

Fill Factor

Fill factor describes how well a test object is electromagnetically coupled to a test coil that surrounds or is inserted into the test object. Fill factor pertains to tests that use inside (bobbin) or outside (encircling) coils.

It is necessary to maintain a constant relationship between the diameter of the coil and the diameter of the test object. Small changes in the diameter of the test object can cause changes in the impedance of the coil. This can be useful in detecting changes in the diameter of the test object, but it can also mask other indications.

The fill factor (Figure 5) is always a number less than 1. It is necessary to have a means to guide the test object through the center of the coil, or vice versa, to maintain a fill factor of constant value.

Fill factor can be described as the ratio of test object diameter squared to the coil diameter squared for an encircling coil, or vice versa for a bobbin coil, in which case the inside diameter is used. As the fill factor

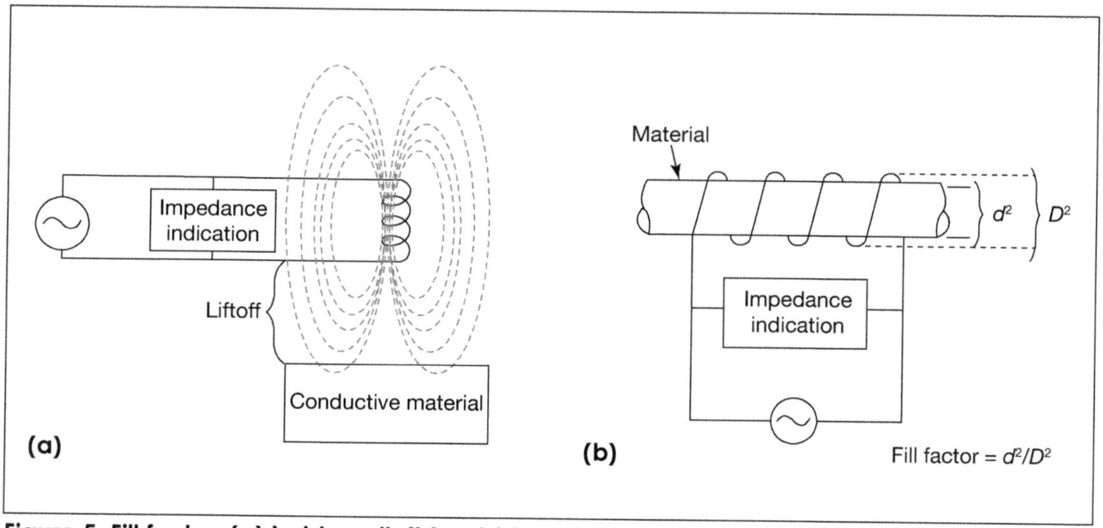

Figure 5. Fill factor: (a) inside coil; (b) outside coil.

approaches 1 the efficiency or the amount of primary flux energy linking the probe to the test material will increase.

(Eq. 2) $$fill\ factor = \frac{d^2}{D^2}$$

Calculation of Fill Factor

Inside coils:

d = coil windings outside diameter
D = tube internal diameter

Outside coils:

d = test object outside diameter
D = coil windings inside diameter

Problem 2

What is the fill factor for inspecting copper tubes with an outside diameter (OD) of 0.75 in. (19 mm) and wall thickness of 0.049 in. (1.24 mm) of using a 0.610 in. (15.5 mm) outside diameter bobbin probe?

Solution

First, calculate the inside diameter (ID) of the copper tubes as follows:

$$D = 0.75 - (0.049 \times 2)$$
$$D = 0.75 - 0.098 = 0.652$$

Use Equation 2 to calculate the fill factor for bobbin coils:

where
d is 0.610
D is 0.652

The solution is calculated as follows:

$$fill\ factor = \frac{(0.610)^2}{(0.652)^2} = \frac{0.372}{0.425} = 0.88$$

Thus, the fill factor is 88%.

Problem 3

What is the coil winding's inside diameter for inspecting 1 in. (25.4 mm) solid rods using a fill factor of 90%?

Solution

Use Equation 2 to calculate the fill factor for encircling coils:

where

fill factor is 90% or 0.9

d is 1

Fill in the variables as follows:

$$0.9 - \frac{(1)^2}{D^2}$$

$$D^2 = \frac{1}{0.9}$$

$$D^2 = 1.111$$

$$D = \sqrt{1.111}$$

$$D = 1.05$$

Thus, the inside diameter of the coil should be 1.05 in. (26.7 mm).

LEVEL II

9

Eddy Current Test Systems and Analysis

Impedance Testing Systems

Impedance testing systems are the simplest to operate. Most of the portable conductivity testers and discontinuity detectors simply detect a change in impedance. They do not detect phase shifts.

Impedance testing, in spite of the many variables that can cause changes in impedance, is actually fairly simple and direct. The most difficult task for the technician is the ability to recognize with any degree of certainty whether a change in impedance is due to a change in liftoff or due to a change in conductivity.

The presence of discontinuities in the material will also cause changes in impedance. Since discontinuities are detected by sudden changes in indications as opposed to gradual changes for all the other conductivity factors, the technician has no difficulty in differentiating the two types.

Phase Analysis Systems

Because a change in impedance is accompanied by a shift in phase, it is possible to observe the phase shifts rather than the impedance change to determine the conditions that exist in the material.

The impedance of a coil may be represented by a vector, whose length represents the impedance value and whose direction represents a phase angle (the angle by which the current lags behind the voltage). These vectors may be measured and plotted on a chart known as the *impedance-plane diagram.* The impedance-plane diagram may be plotted either by knowing the impedance value and the phase angle or by knowing the resistive component and the reactive component.

Conductivity on the Impedance-Plane Diagram

Figure 1 is a typical impedance-plane representation of several impedances obtained by eddy current testing from several different types of nonferromagnetic metals with the same test set and at the same frequency with different conductivity. The lag angles and impedance values obtained for each of these materials, including the values obtained when the coil was held in air, are shown by their individual vectors.

In Figure 2, a curve is drawn connecting the impedance value of each of the vectors. This curve then is the locus of all the impedances that will result from changes in conductivity. If all other factors are held constant, a change in conductivity of any material will result in an impedance

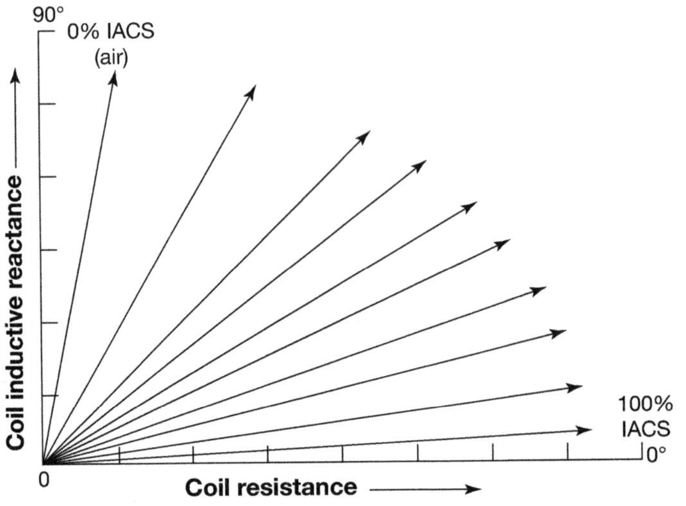

Figure 1. Impedance vectors shown on the impedance-plane diagram.

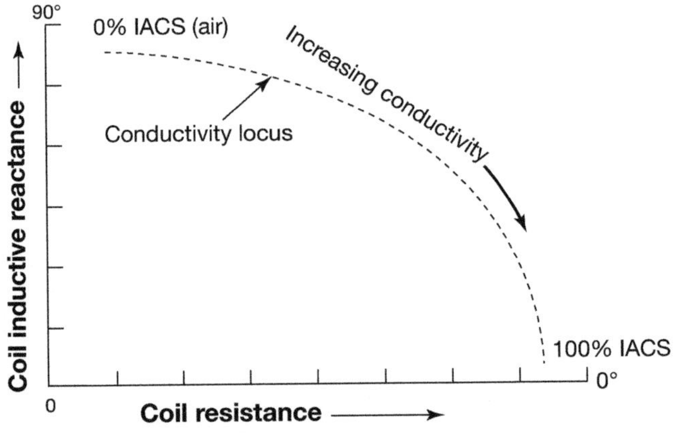

Figure 2. Conductivity locus on the impedance-plane diagram for nonferromagnetic materials.

value that will fall somewhere on this curve. An increase in conductivity will cause the impedance to move to the right and down along the curve, while a decrease in conductivity causes the impedance to move up and to the left.

Figure 3 plots the locus of all the impedances that result as the coil is lifted off the 100% International Annealed Copper Standard (IACS) material (all other factors held constant). When the coil is held in contact with the 100% IACS material, the locus is on the conductivity curve. As the coil is lifted off the material, the impedance moves in the direction indicated by the dashed line. As the coil is further removed, the impedance moves further along the dashed line until the material no longer affects the coil (the coil is in air). At this point, the impedance is back on the conductivity curve at 0% IACS, as shown.

Figure 4 illustrates a family of liftoff curves, each one starting at a different conductivity. Note the angles (*A* and *B*) formed at the junctions of the liftoff locus curves and the conductivity locus curve. In the areas of low conductivity on the diagram, angle *B* is small. In the

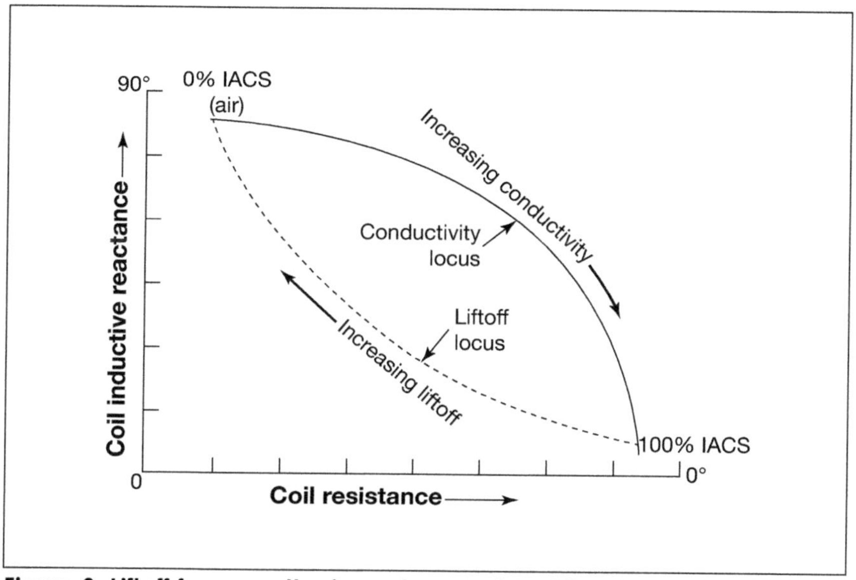

Figure 3. Liftoff focus on the impedance-plane diagram.

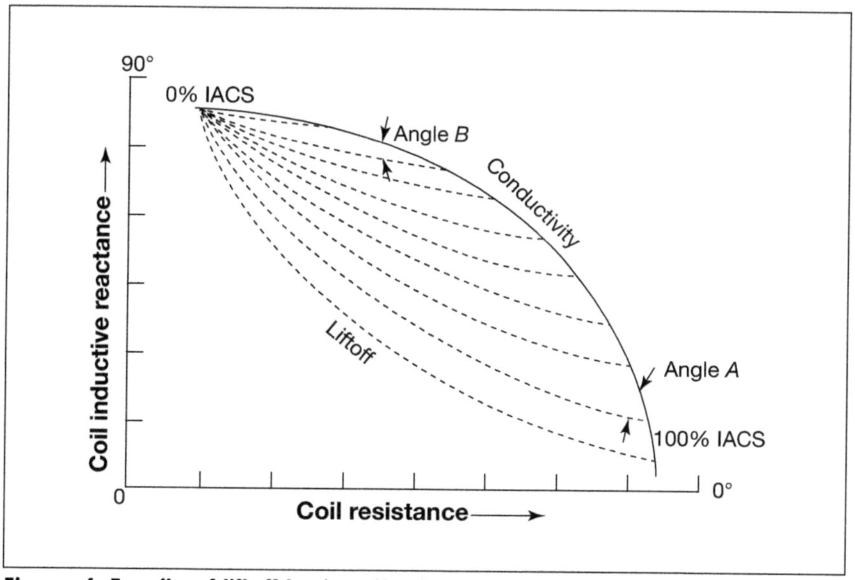

Figure 4. Family of liftoff loci on the impedance-plane diagram.

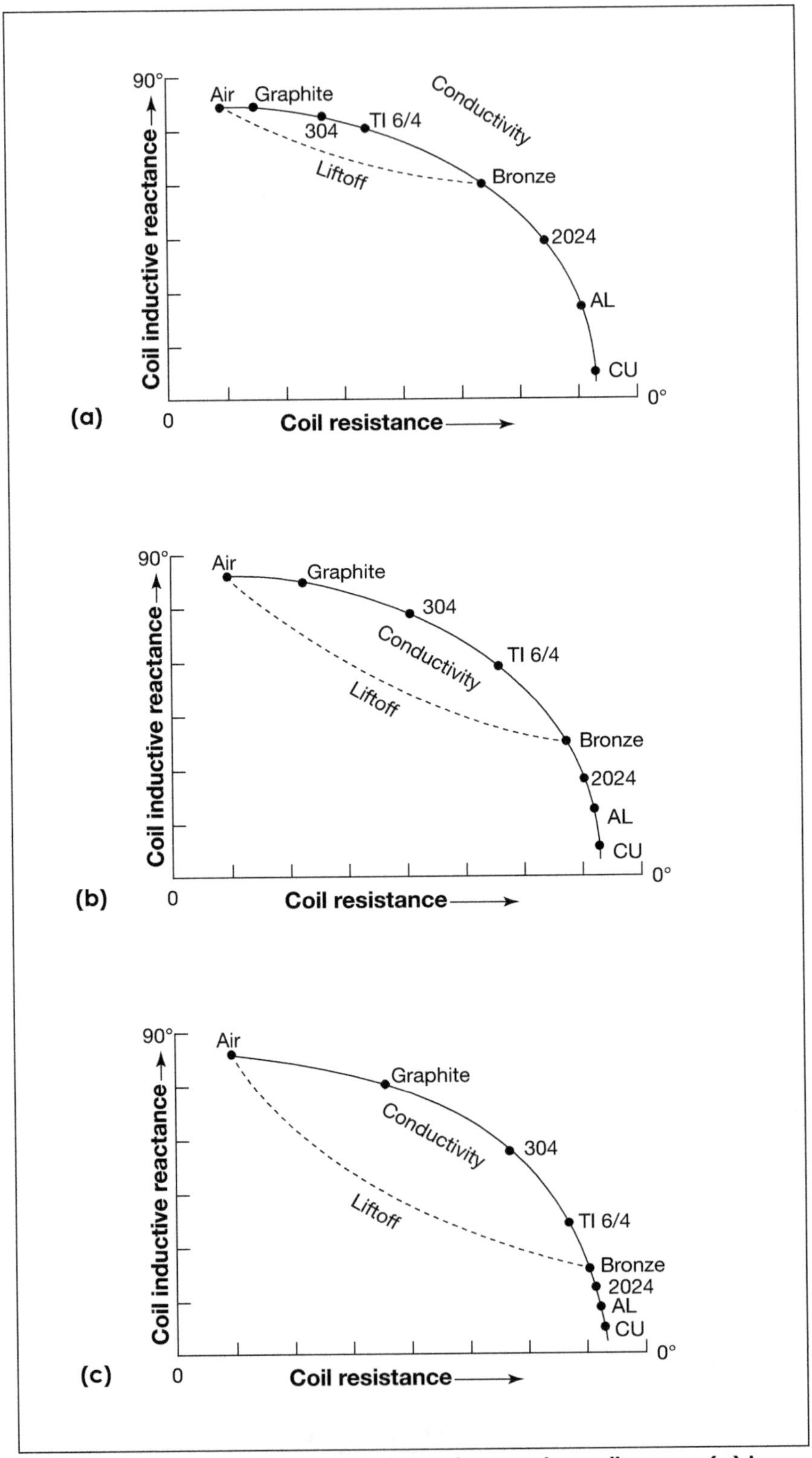

Figure 5. Frequency effect on the impedance-plane diagram: (a) low frequency (20 kHz); (b) medium frequency (100 kHz); and (c) high frequency (1 MHz).

areas of high conductivity, angle *A* is larger. Thus, the impedance-plane diagram shows that in materials with high conductivity, it is easier to detect a change in liftoff. In areas of high conductivity on the diagram, a change in conductivity is mostly a change in the coil's inductive reactance (vertical), while changes in liftoff are mostly a change in the coil's resistance (horizontal). Due to this large difference in direction, the effect of liftoff may be distinguished from the effect of a change in conductivity.

Effect of Frequency on the Impedance-Plane Diagram

If the frequency of the voltage applied to the coil is changed, there is a change in the resistive and reactive factors of the impedance. If an impedance-plane diagram of conductivity is then prepared at the new frequency, the conductivity locus curve will be different than that obtained at the old frequency. Although the specific values are different, the two curves are very similar in appearance. Figure 5 shows three typical conductivity loci that might be obtained at three different operating frequencies.

Note that at higher frequencies, the locus of the point representing any given material—in this case, bronze—shifts toward the lower end of the conductivity curve. Also note that at lower test frequencies, copper and aluminum are widely separated on the conductivity loci curve, while at higher test frequencies, they are much closer together on the lower end of the curve.

Effect of Material Thickness

Figure 6 shows the effect of thickness changes as applied to brass. The thickness loci may be plotted by measuring the resistance and reactance of a very thin sheet of brass and repeating the process for steadily increasing thicknesses. Ultimately the depth of penetration limit is equaled and the resistance and reactance values reach the conductivity curve. Additional thickness increases will have no effect.

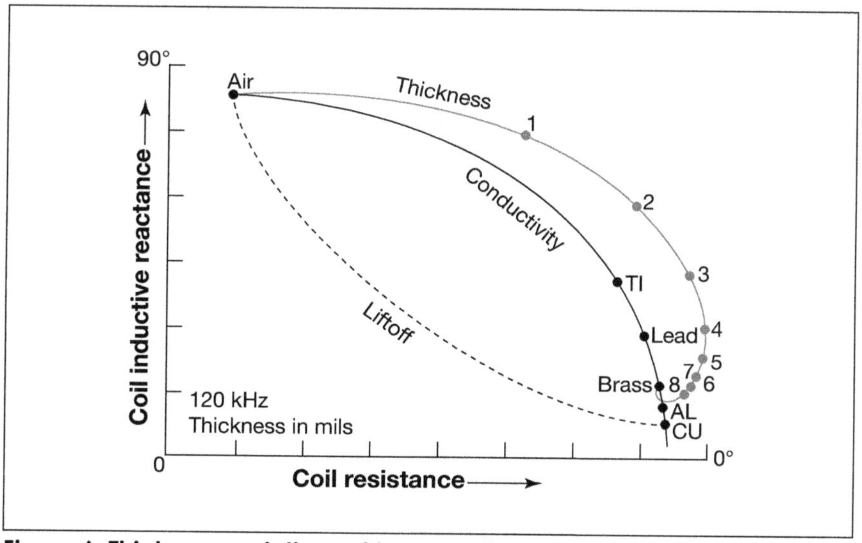

Figure 6. Thickness variations of brass on the impedance-plane diagram.

Figure 7 shows thickness loci for several different types of metals. Since the eddy currents gradually become weaker and weaker as they approach the depth of penetration limit, thickness variations have less and less effect (note how the distance between points on the curve become shorter for the same change in thickness as the material thickness is increased). Hence, the sensitivity to varying thicknesses decreases as the material thickness increases.

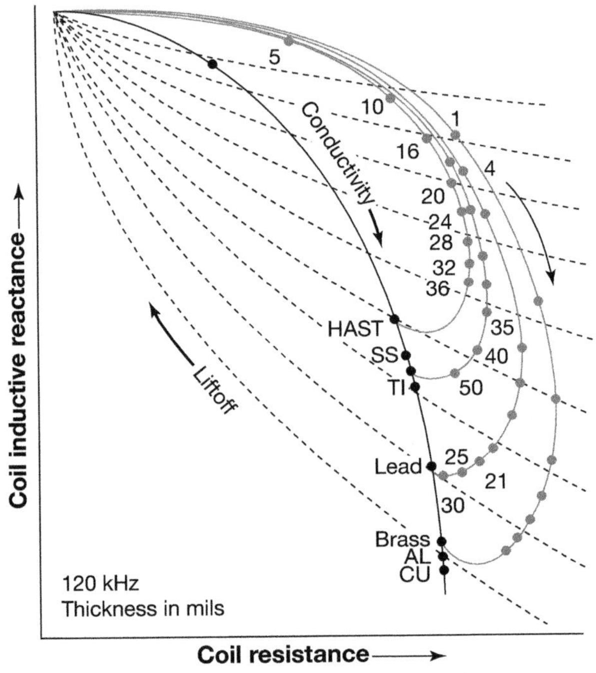

Figure 7. Thickness variations of several metals on the impedance-plane diagram.

Effect of Frequency on Thickness Measurements

Figure 8 shows the effect that a change in frequency has on thickness measurements of brass. Since a change in frequency changes the depth of penetration, a lower frequency will give a greater depth of penetration when the conductivity curve is reached. The thickness points that are plotted will be located higher on the thickness curve with more distance between them. Raising the frequency will have the opposite effect. The depth of penetration will not be as great as the points that are plotted. These will be located farther down on the curve and closer together.

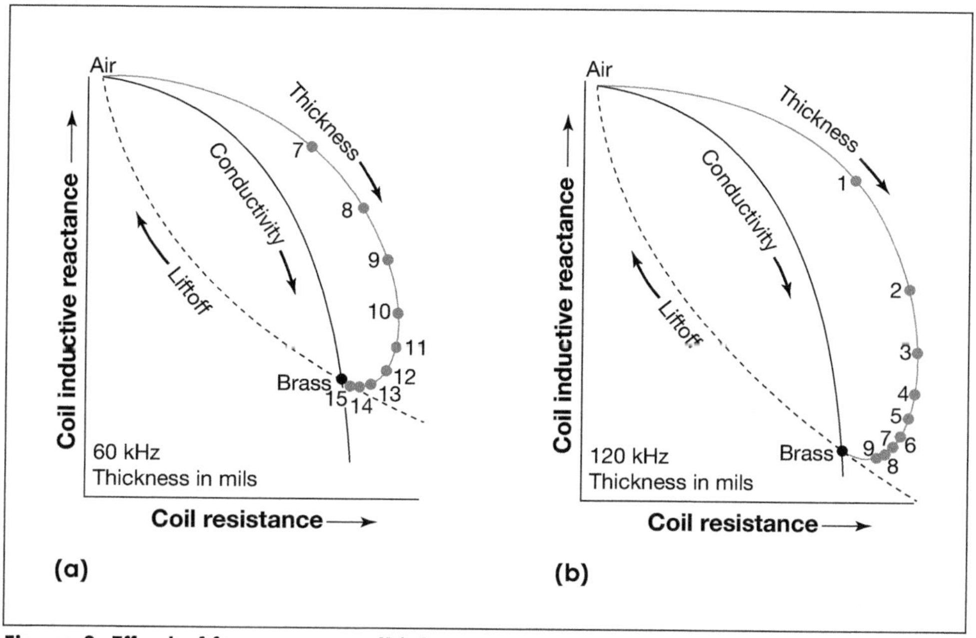

Figure 8. Effect of frequency on thickness measurements (1 mil = 0.001 in. = 0.0254 mm): (a) 60 kHz; (b) 120 kHz.

Suppression of Nonrelevant Variables

In the type of eddy current equipment where the bridge networks may be balanced (or nulled) by adjustment of resistance and reactance, the bridge may be set to operate from any point on the impedance-plane. Figure 9 illustrates the basic concept of this approach.

Suppression of Liftoff Variable

Figure 9a shows the selection of test points that could be used to suppress the liftoff variable. Suppression of the liftoff variable is accomplished by simply selecting a test point D so that the distances D–C and D–A are equal. Thus, a slight change in liftoff will not affect the overall reading.

The test point selected could lie anywhere along the line D–E. If the test point selected is too far away from the conductivity curve, it is possible that the meter will be driven off scale. Selecting a test point considerably away from the conductivity curve will produce the best overall results.

If test point D is selected as the test point and the probe is moved from alloy A to alloy B, the meter reading will decrease. If point E is selected as the test point, the meter reading will increase when the probe is moved from alloy A to alloy B. It is possible that better sensitivity can be obtained from a test point located on one side than on another. In practice, the best procedure is to evaluate test points on both sides, then choose whichever gives the best results.

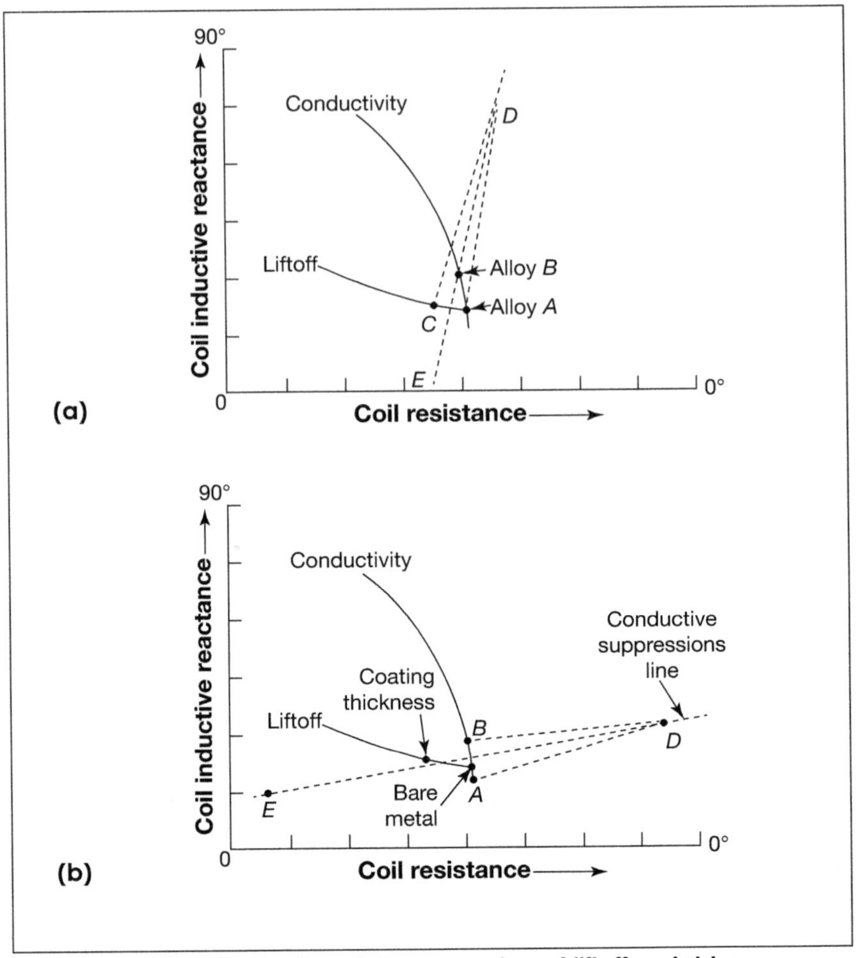

Figure 9. Test point selection: (a) suppression of liftoff variable; (b) suppression of conductivity variable.

Suppression of Conductivity Variable

In measuring the thickness of nonconductive coatings on conductive materials, it is necessary that the liftoff variable be measured and that the conductivity variable be suppressed. To accomplish this, a test point is selected, as shown in Figure 9b, which lies on a line that is perpendicular to the conductivity curve at the point of interest.

Conductivity and Permeability

Figure 10 shows that the reactance component of the test coil impedance is decreased by the presence of nonferromagnetic materials. This reactance reduction occurs because induced currents flow in the conductive and nonferromagnetic object and set up a secondary field that partially cancels the primary field of the coil. The opposite is true when a ferromagnetic material such as iron or ferrite is placed within the field of the coil. This happens because the presence of the magnetic field intensity of the primary coil field causes atomic magnetic elements of the ferromagnetic material to become aligned with the field, increasing the

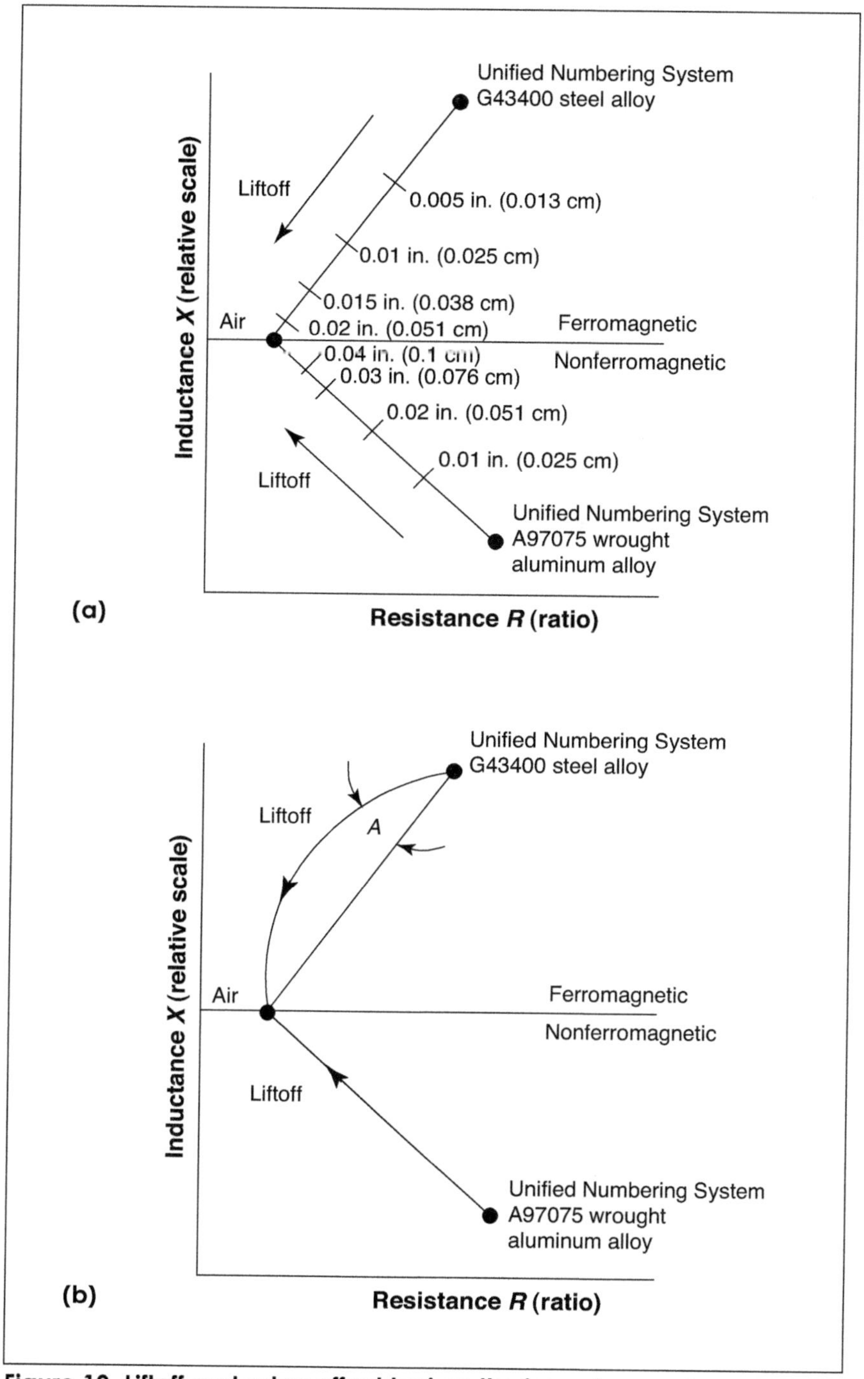

Figure 10. Liftoff and edge effect loci on the impedance plane: (a) liftoff loci; (b) edge effect loci.

flux density. As stated previously, the magnetic permeability (μ) is the ratio of flux density (*B*) to magnetic field intensity (*H*).

For ferromagnetic materials, as shown in Figure 11a, the liftoff and magnetic permeability loci curves are virtually superimposed, but their respective values increase in opposite directions. The nickel-zinc ferrite cores in Figure 11b were chosen as examples because they have a low conductivity and two different values for permeability. The effect of the increased flux density gives a greater induced voltage in the test coil that in turn raises the impedance. The increase in impedance is in the reactance direction except for the effect of a small amount of energy loss resulting from hysteresis. The nickel-zinc ferrite cores may have an initial permeability of 850, which is high on the permeability line. Usually, practical engineering materials also have an associated electrical conductivity that affects the impedance, as shown for 422 steel and 4340 steel. The relative relationship of the permeability loci lines and the conductivity curves for three materials is shown in Figure 11c.

The vector or phasor values of inductive reactance and resistance for different material conditions yield unique loci or phasor plots on the impedance plane at particular operating frequencies. The phase angle of the impedance vectors will change at different frequencies because the inductive reactance value is a function of inductance and

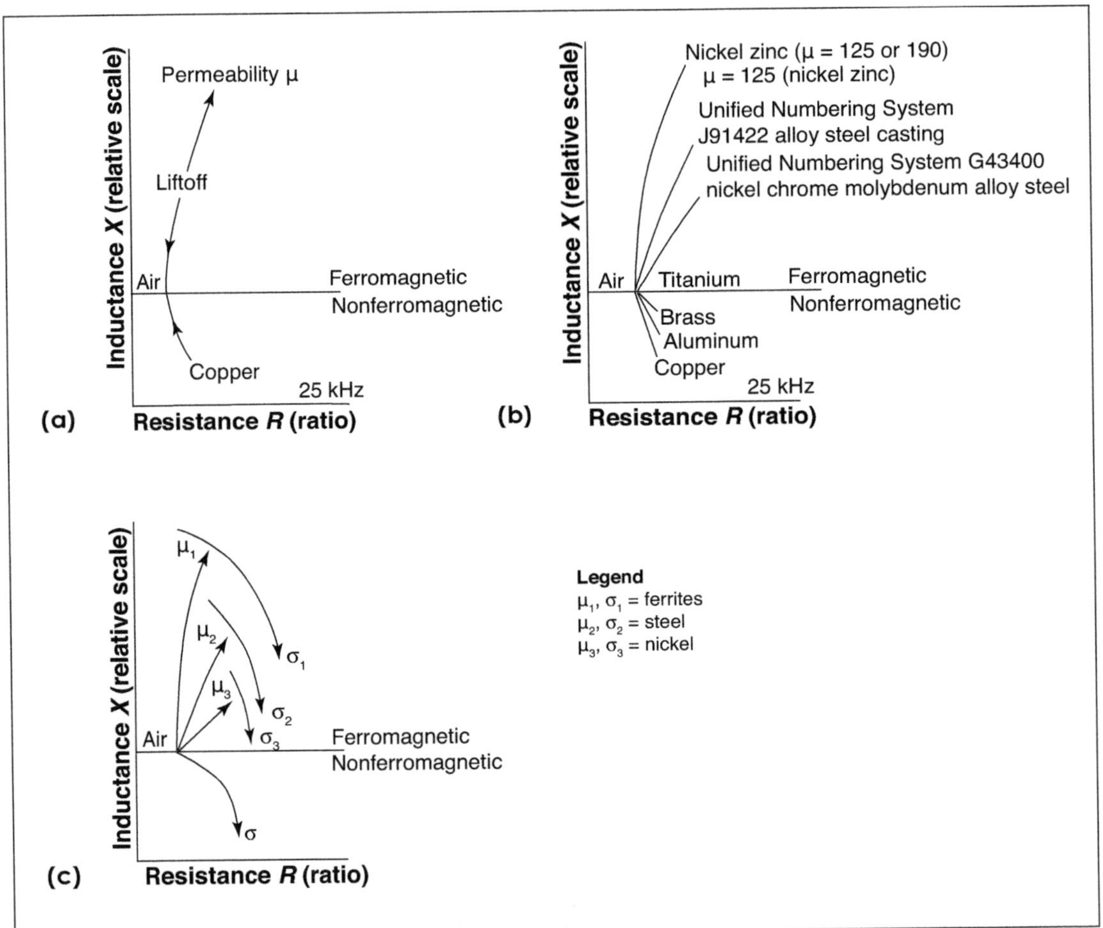

Figure 11. Permeability, liftoff, and conductivity loci on the impedance plane: (a) permeability and liftoff locus; (b) permeability loci for different materials; and (c) loci for permeability (μ) and conductivity (σ).

frequency. Hence vector points may move relative to one another along the conditional loci curves when the operating frequency is changed. This shift in phase is shown in Figure 12 for the conductivity values of nonferromagnetic materials.

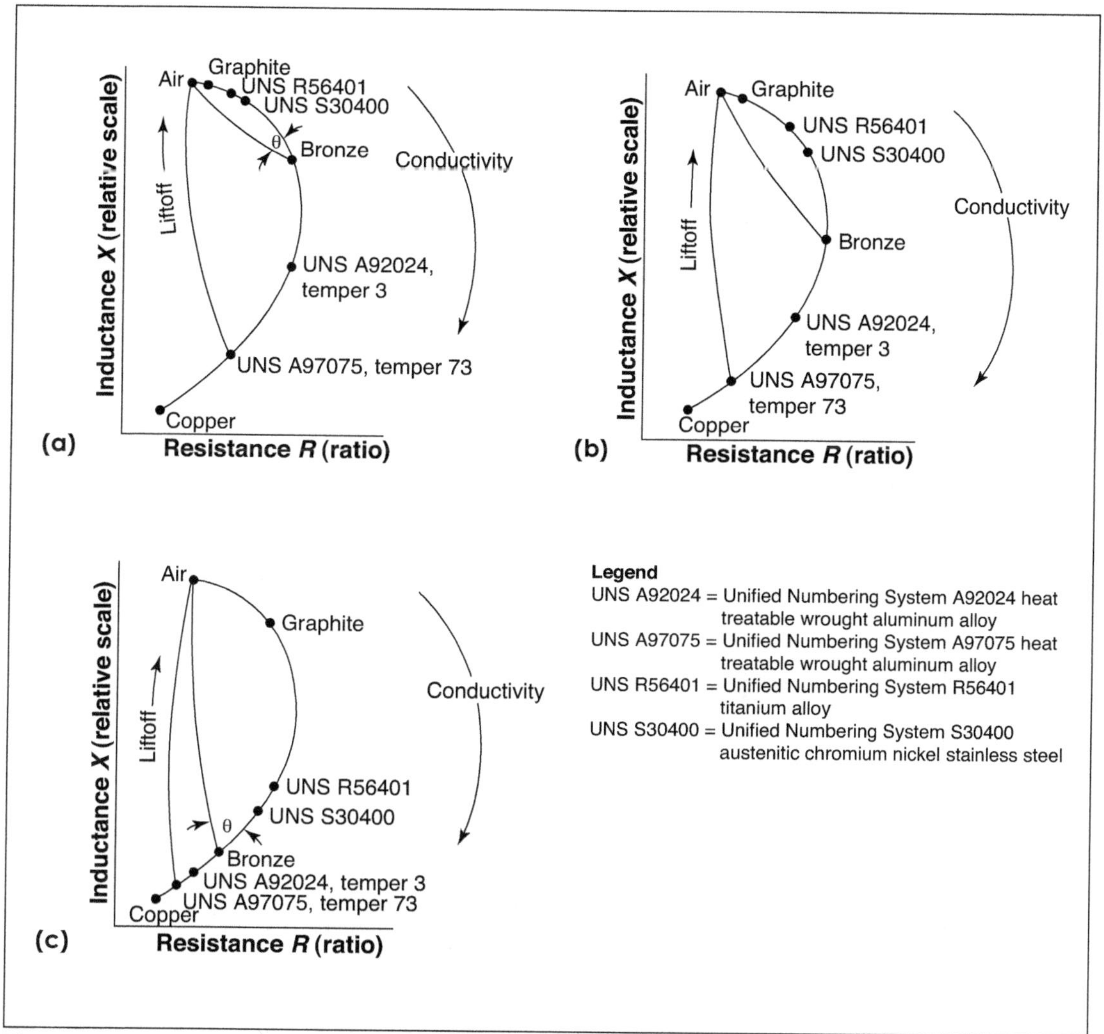

Figure 12. Movement of material points by frequency changes: (a) low frequency (20 kHz); (b) medium frequency (100 kHz); and (c) high frequency (1 MHz).

Similar phase angle changes for the permeability of 4340 steel are shown in Figure 13 as the frequency changes from 75 to 300 kHz. These changes in phase shift at different frequencies do not interfere with impedance-plane analysis, provided that the operator is aware of this factor. In some cases, test results may be improved by changing the frequency to cause phase shifts.

With phase-analysis eddy current instruments, an operator can produce impedance-plane loci plots or curves automatically on a *flying dot* display (see "Vector Point Method" in the following section). Such

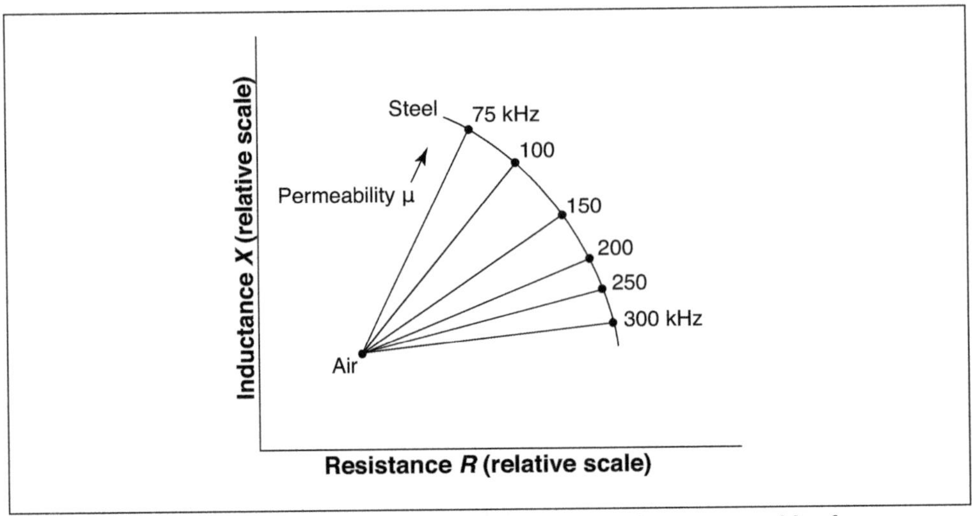

Figure 13. Phase angle changes on the impedance-plane caused by frequency changes.

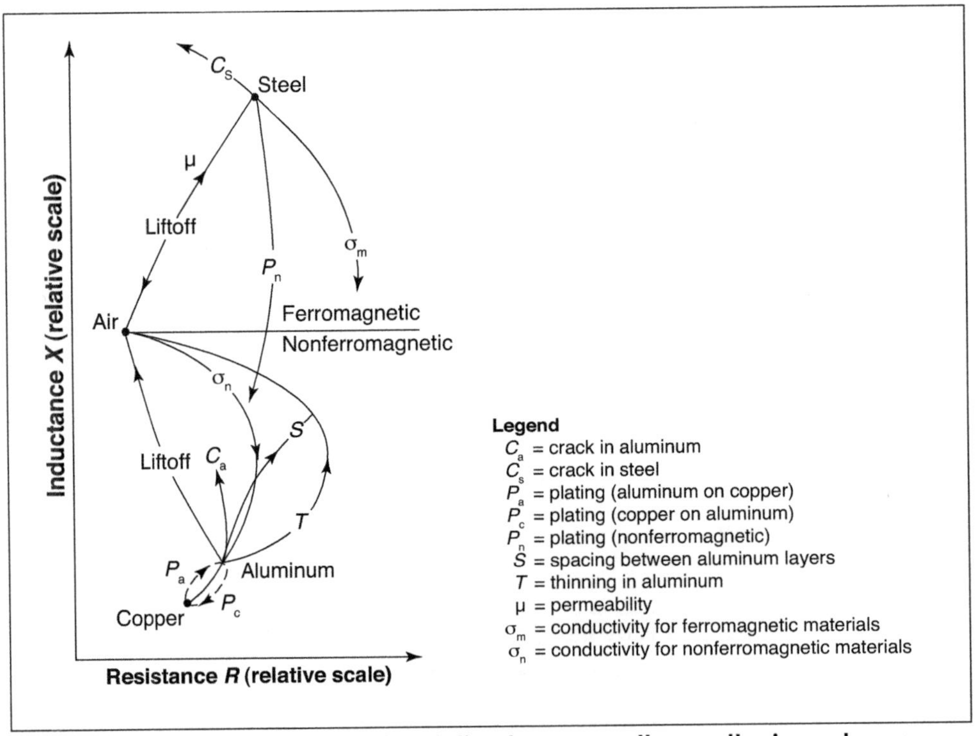

Figure 14. Impedance changes in relation to one another on the impedance plane.

impedance-plane plots can be presented for the following material conditions, as shown in Figure 14:

- Liftoff and edge effects
- Cracks
- Material separation and spacing
- Permeability

- Specimen thinning
- Conductivity
- Plating thickness

Evaluation of these plots shows that ferromagnetic material conditions produce higher values of inductive reactance than values obtained from nonferromagnetic material conditions. Hence, the ferromagnetic domain is in the upper quadrant of the impedance plane, whereas nonferromagnetic materials are in the lower quadrant. The separation of the two domains occurs at the inductive reactance values obtained with the coil removed from the conductor (sample). This is proportional to the value of the coil's self-inductance, *L*.

Linear material values do not produce linear responses on the impedance-plane loci. When the eddy current probe is balanced on the metal specimen, the loci values for linear material conditions are displayed as follows:

- Ferromagnetic and nonferromagnetic liftoff conditions are displayed logarithmically (in X).
- Ferromagnetic and nonferromagnetic edge effects are displayed logarithmically.
- Ferromagnetic and nonferromagnetic conductivities vary with test frequency.
- Magnetic permeability varies with test frequency.
- Metal thinning varies exponentially.
- Nonferromagnetic plating thickness is displayed logarithmically.
- Material spacing or separation varies exponentially.

In a problem-solving situation, impedance-plane analysis is a useful tool because it improves the ability to detect various conditions and provides a better understanding and interpretation of the eddy current test results.

Digital Displays

Vector Point Method

When a digital display is provided as part of the test equipment, the equipment may be set up to show on the screen the locus of all the points of interest to the technician. Thus, the operator may construct, point by point, the impedance-plane diagram directly on the screen. Digital systems allow the setup of standardization curves constructed from known standards and the automated analysis of signals as compared to these curves. During an actual eddy current test, the impedance of the coil causes a dot, referred to as a flying dot, to appear at some point on the screen. Its position with respect to the impedance-plane diagram tells the technician what is occurring within the test object. Essentially, the point represented by the flying dot describes the in-phase and quadrature components of the signal.

An advantage of using a digital display is its flexibility. For example, the equipment may be set up so that the display is rotated to a position where a change in liftoff would move the dot left or right, while a change in conductivity would move the dot up or down, as shown in Figure 15.

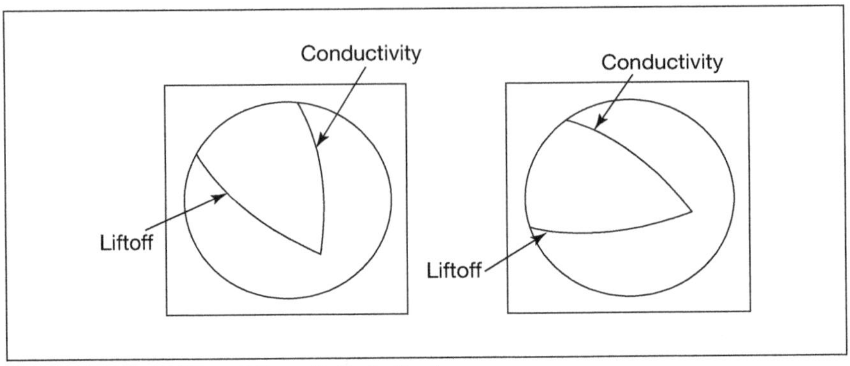

Figure 15. Rotation of the digital display.

Ellipse Display Method

A digital display may also be set up to compare a test object with a reference standard. The ellipse method uses an inspection coil in conjunction with a reference coil. When a standard is placed under the reference coil and the test object is placed under the inspection coil, the display shows the phase relationships between the signals obtained. This comparison between the signals provides indications of the dimension variable and the conductivity variable.

The dimension variable and the conductivity variable are shown on the display by the width of the ellipse and the angle tilt of the axis. Figure 16 shows the appearance of ellipses for dimension and

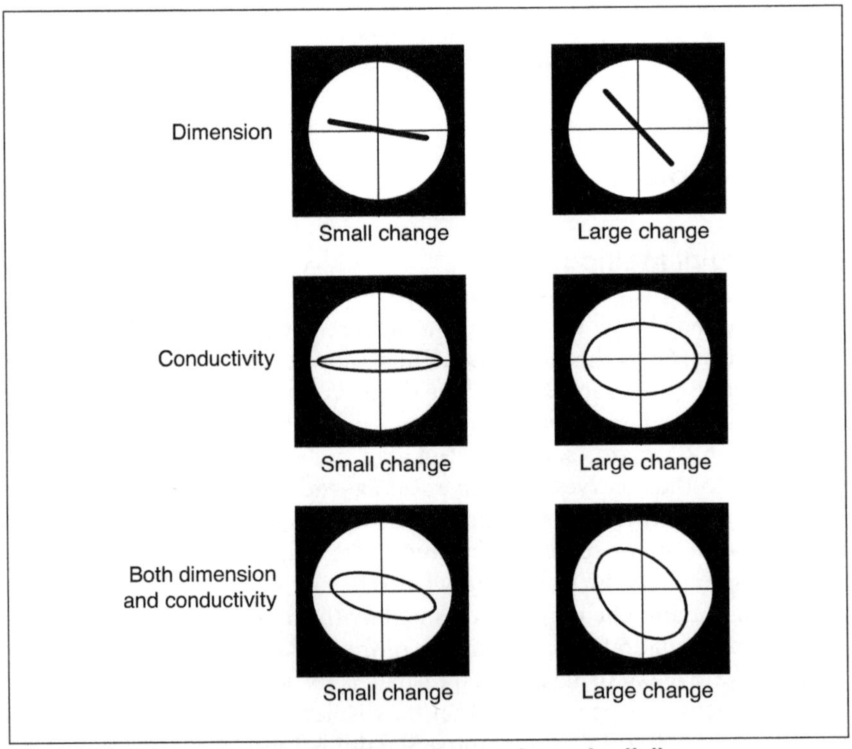

Figure 16. Ellipse displays for dimension and conductivity.

conductivity. Figure 17a shows an X-Y screen display of the single 0.031 in. (0.787 mm) hole drilled through the weld seam of an aluminum tube shown in Figure 17b. The sample was used to optimize instrument settings. A test frequency of 15 kHz provided good phase separation of the noise and discontinuity signals. The indication from an inherent discontinuity in the form of a scratch is labeled on the screen as an imperfection.

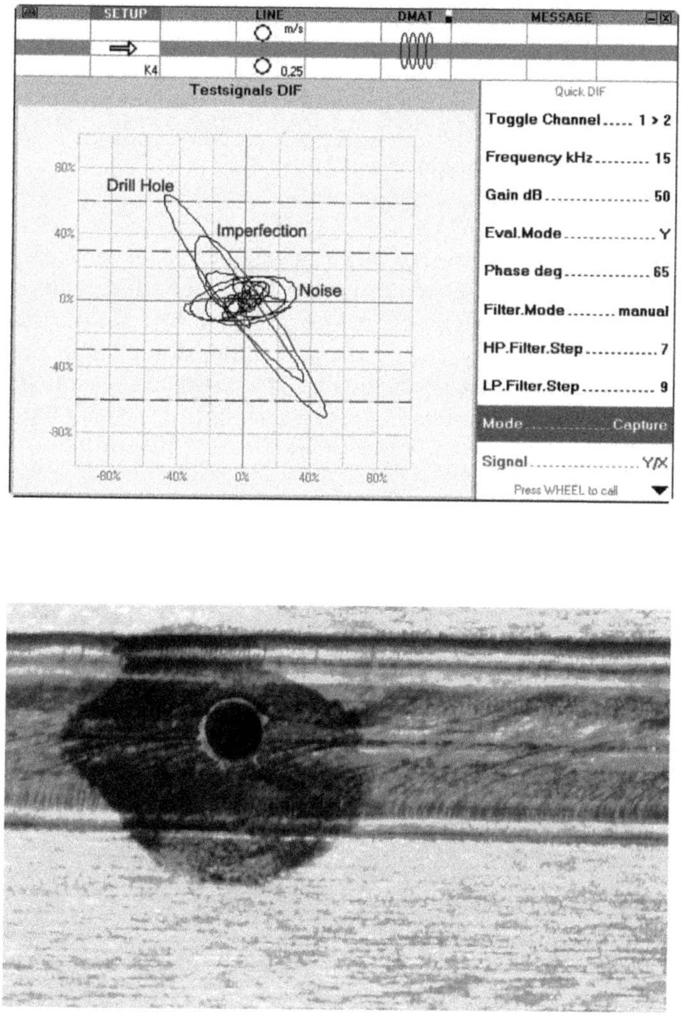

(a)

(b)

Figure 17. Electromagnetic test of an artificial discontinuity: (a) X-Y screen display; (b) test sample with drilled hole.

Modulation Analysis

A modulation analysis system, shown schematically in Figure 18, adds a modulating device between the test set and an indicating device or recorder. The modulating device is simply an electronic filter that will pass only certain frequencies. In modulation analysis, a differential coil is used so that two adjacent areas of the test object are compared.

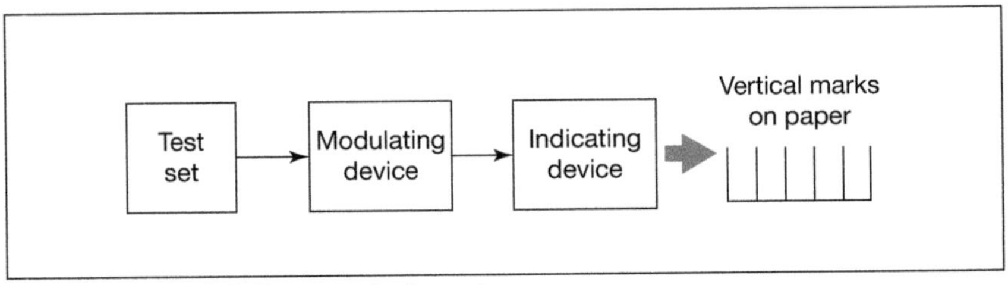

Figure 18. Modulation analysis system.

As the test object passes under or through the coils at a constant rate, various variables being sensed cause the equipment to register a signal. A discontinuity such as a crack will be indicated by a sharp rise in signal, followed immediately by a sharp drop. A dimension change, on the other hand, is most likely to occur gradually. Thus, with the use of either a low-frequency or high-frequency filter, the effect of one variable or the other is eliminated from the readout, as shown in Figure 19.

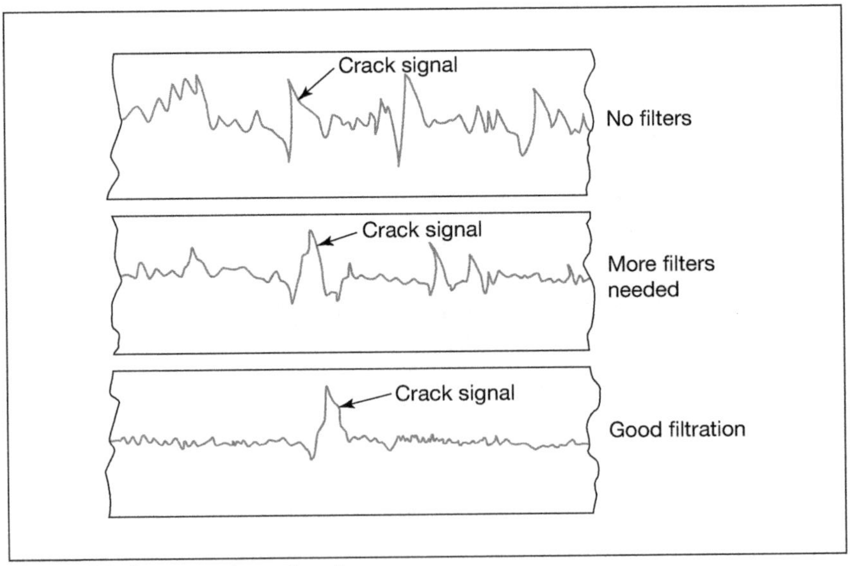

Figure 19. Suppression of noise.

10

Selection of Test Frequency

Test Frequency

The frequency of an alternating current is defined as the number of cycles (one complete cycle of current) that occur in 1 s. Its unit is the hertz, one hertz (1 Hz) being one cycle per second. Thus, house current at 60 cycles per second has a frequency of 60 Hz. The most important parameter of the test system that affects the depth of penetration is the test frequency.

The type of alloy involved and the variables to be measured or suppressed determine the best frequency. For a given alloy, higher frequencies normally limit the eddy current test to inspection of the excited metal surface nearest the primary coil winding. Lower frequencies permit deeper eddy current penetration. At any selected test frequency, eddy currents will penetrate deeper in lower conductivity alloys than in higher conductivity alloys.

High test frequencies are normally used for detecting small surface cracks or surface contamination and for gauging thin coatings. Medium frequencies are useful for conductivity measurements such as alloy sorting. Low test frequencies are usually required for testing thicker materials (for opposite side corrosion, for example), for thickness gauging, and for penetrating into ferromagnetic materials.

Penetration depth, however, is only part of the process for selection of optimum eddy current test frequencies. The geometric relationship between the impedance curves for the variable magnitudes (different conductivity points or liftoff points) along their respective impedance curves is important.

Depth of Penetration

Eddy currents are not uniformly distributed throughout a test object. They are most dense at the surface closest to the coil and become progressively less dense with increasing distance below the surface of the material. At some distance below the surface of a thick material there will be essentially no currents flowing. The depth of penetration is affected by the frequency, conductivity, and permeability of the material.

- The depth of penetration decreases as the frequency increases.
- The higher the conductivity, the less the penetration.
- The higher the permeability, the less the penetration.

When testing ferromagnetic materials, the permeability factor will have no effect on the depth of penetration if the test object is magnetized to saturation by a separate direct current coil. The standard depth of penetration is defined as the depth at which the current strength has dropped to 37% of the current density that exists at the surface. Figure 1 shows the distribution of eddy currents in a material.

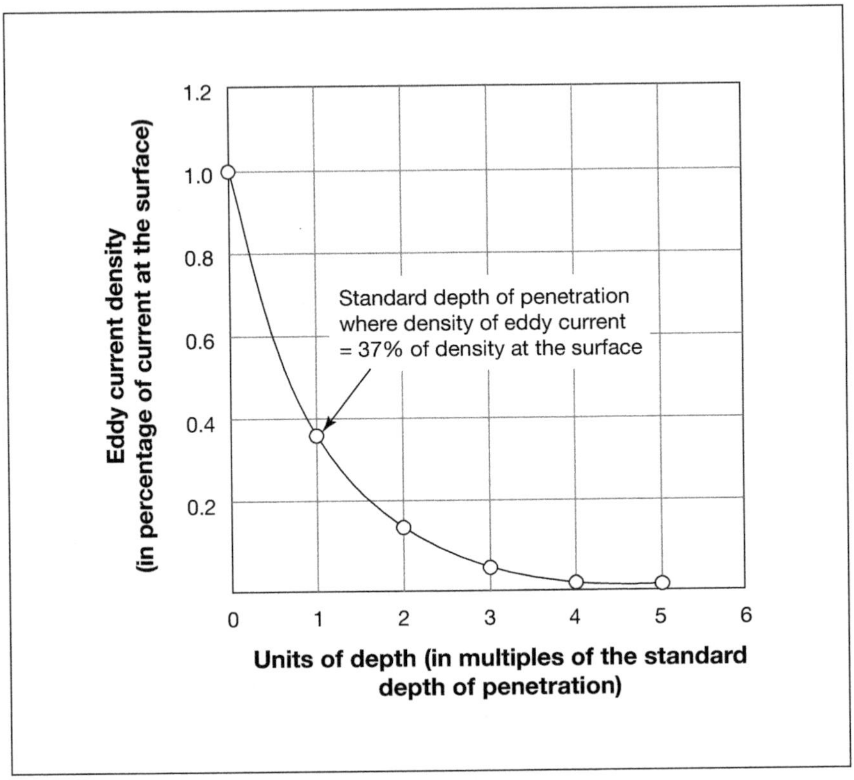

Figure 1. Variation in eddy current density.

If the standard depth of penetration exceeds the thickness of the material under test, the restriction of the eddy current paths appears as a change in the conductivity of the material. The coil response then reflects the thickness of the material. Eddy currents do not cease to exist beyond one standard depth. Normally, the material must have a thickness of two or three times the standard depth before thickness ceases to have any effect on the test coil. Figure 2 shows the standard depth of penetration for several materials with different conductivities at various operating frequencies.

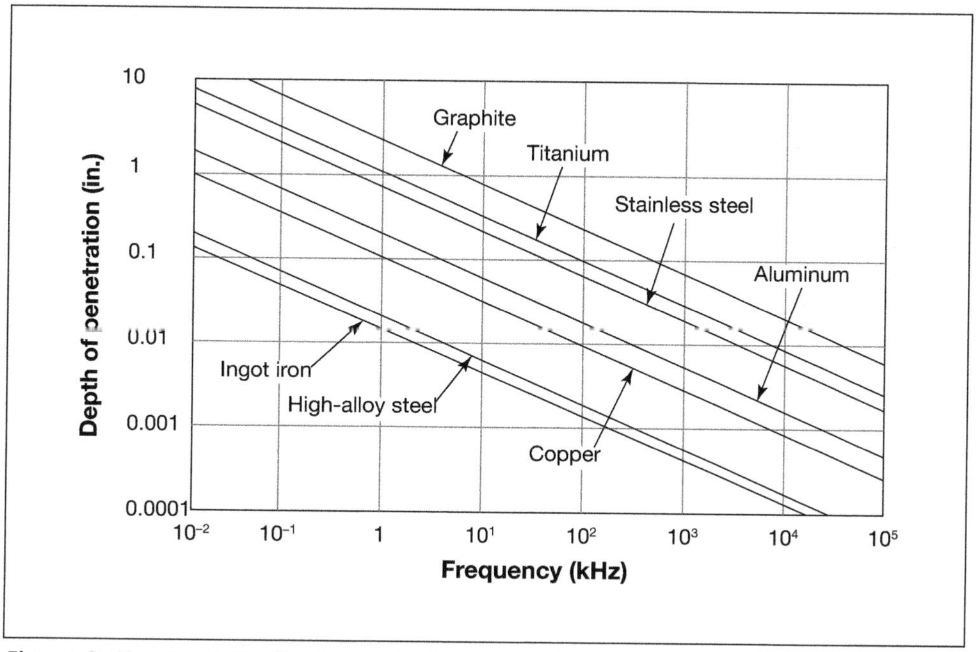

Figure 2. Standard depth of penetration versus frequency for different materials.

The standard depth of penetration (δ) is calculated as follows:

(Eq. 1)

$$\delta = \frac{1}{\sqrt{\pi f \mu \sigma}}$$

where

- δ is standard depth of penetration (m)
- π is 3.1416
- f is frequency (Hz)
- μ is magnetic permeability (H/m)
- σ is electrical conductivity (S/m)

Problem

A stainless steel block (300 series) is tested with a surface probe operating at a frequency of 100 kHz. Stainless steel (300 series) is nonferromagnetic. Its magnetic permeability (μ) is $4\pi \times 10^{-7}$ H/m. Its conductivity is 0.14×10^7 siemens (mhos) per meter. What is the standard depth of penetration in millimeters for eddy currents in the block?

Solution

Given
$\pi = 3.1416$
frequency (f) = 100×10^3 Hz
magnetic permeability (μ) = $4\pi \times 10^{-7}$ H/m
conductivity (σ) = 0.14×10^7 S/m

Use the formula for calculating depth of penetration and plug in the numbers to calculate the answer as follows:

$$\delta = \frac{1}{\sqrt{3.1416 \times (100 \times 10^3) \times (4 \times 3.1416 \times 10^{-7}) \times (0.14 \times 10^7)}}$$

$$\delta = \frac{1}{\sqrt{3.1416 \times 100\ 000 \times (12.5664 \times 0.0000001) \times (0.14 \times 10\ 000\ 000)}}$$

$$\delta = \frac{1}{\sqrt{552\ 700.43}}$$

$$\delta = \frac{1}{743.438}$$

$$\delta = 0.00135 \text{ m}$$

$$\delta = 1.35 \text{ mm}$$

Thus, the depth of penetration for this particular situation is 1.35 mm (0.05 in.).

Limit Frequency

An equation for determining the limit frequency (f_g) is:

(Eq. 2)
$$f_g = \frac{5066}{\mu_{rel} \sigma d^2}$$

where
f_g is limit frequency (Hz)
μ_{rel} is relative permeability
d is diameter of test object (cm)
σ is conductivity (m/Ω-mm^2)

Limit frequency (f_g) is then established in terms of conductivity, permeability, dimension, and a constant (5066). Since limit frequency is based on these parameters, frequency determination using a test frequency to limit frequency ratio f/f_g can be calculated. High f/f_g ratios are used for near-surface tests, and lower f/f_g ratios are used for subsurface tests.

Single-Frequency Systems

Single-frequency systems are capable of energizing the test coil with a single frequency. Frequency selection often becomes a compromise. It is common practice in in-service inspection of thin-wall, nonferromagnetic tubing to establish a standard depth of penetration just past the midpoint of the tube wall. This permits approximately 25% of the available eddy current to flow at the outside surface of the tube wall. In addition, this establishes a phase difference of approximately 150 to 170° between the inside and outside surface of the tube wall. The combination of 25% outside, or surface current, and a 170° included phase angle provides good detectability and resolution for thin-wall tube inspection.

Multifrequency Systems

Test systems driven by more than one frequency are called *multifrequency* or *multi-parameter systems*. It is common for a test coil to be driven with three or more frequencies. Although several frequencies may be applied simultaneously or sequentially to a test coil, each of the individual frequencies follows rules established by single-frequency methods. Signals generated at the various frequencies are often combined or mixed in electronic circuits that algebraically add or subtract signals to obtain the desired result. The technician must have a good working knowledge of current density and phase relationships in order to make intelligent frequency choices.

Modern multifrequency eddy current systems allow the technician to perform discontinuity detection using the differential mode and absolute mode simultaneously. These systems generally use the time-sharing technique to allow testing at four frequencies at the same time. Each frequency channel is capable of generating differential or absolute data.

The displays on modern test systems allow the operator to monitor multiple channels in real time. These channels may contain raw data in absolute or differential modes as well as filtered or mixed data channels. This allows the interpreter to evaluate all information available from the examination at the same time.

11

Eddy Current Testing Applications

Eddy Current Applications

Applications of eddy current testing in industry are numerous and widespread. Eddy current testing can detect discontinuities that lie in planes transverse to the eddy currents present in the material or detect thickness changes and measure nonconductive coating thicknesses. Discontinuities include cracks, seams, laps, pits, and laminations at cut edges of sheet or plate. The presence of a discontinuity is characterized by a sudden, brief change in impedance as the test coil moves across the discontinuity. Readings are standardized with reference standards having known discontinuities, either natural or manufactured, in the same kind of material.

Eddy current general surface scanning can be done in almost all market sectors (Figure 1). The inspection techniques used are driven by two factors:

- The location of the discontinuities of interest.
- The chemical makeup of the test specimens.

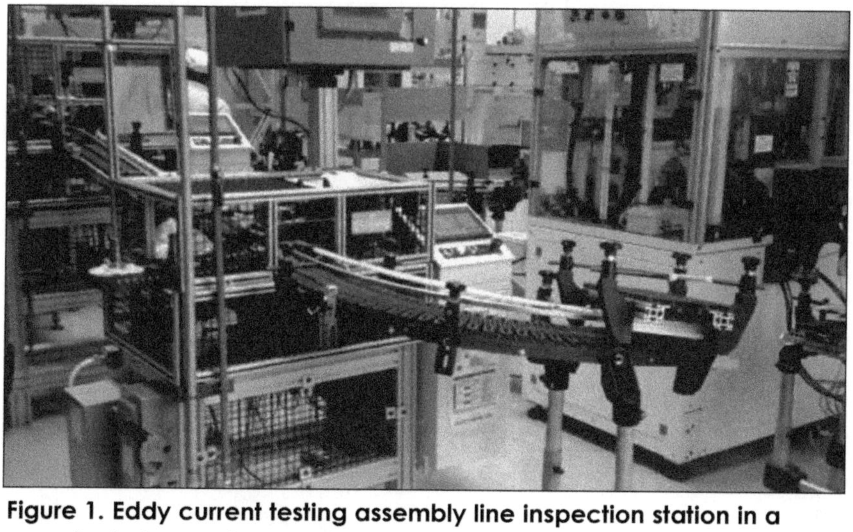

Figure 1. Eddy current testing assembly line inspection station in a manufacturing environment.

If the discontinuities are the result of a general cracking mechanism, they are most likely to occur at the probe inspection point (open to the near surface). In this case, it is possible to inspect both ferritic and nonferritic alloys (Table 1). If the alloys are ferritic and the discontinuities are below the inspection interface, the use of eddy current testing is probably not a good option. Most surface-scanning probes lack the ability to detect subsurface discontinuities in ferritic alloys due to permeability issues. The coil's magnetic field cannot propagate into the test material beyond the first crystalline layer.

Table 1. Discontinuities in ferritic and nonferritic alloys detectable by eddy current testing

Material	Discontinuities	Eddy current testing instruments
Ferritic alloys	Near-surface cracks	Special optimized probe coils
Nonferritic alloys	Near-surface cracks	Small-diameter high-frequency (HF) pancake or pencil probes
	Far-surface or subsurface cracks	Larger-diameter low-frequency (LF) spot or driver/pickup probes
Welds	Cracks in ferritic and nonferritic alloys	Special probe coils optimized for weld inspection

Weld inspection probes have a special coil arrangement that decreases their sensitivity to both permeability variations (due to the material's magnetic properties) as well as liftoff response (due to changing geometry). If properly designed and used, these special probes should still have good signal-to-noise ratios (SNRs) from crack-like events (local conductivity variations) as long as they are open to the surface. A comparison of eddy current signals from a ferritic plate and ferritic weld crown is shown in Figure 2.

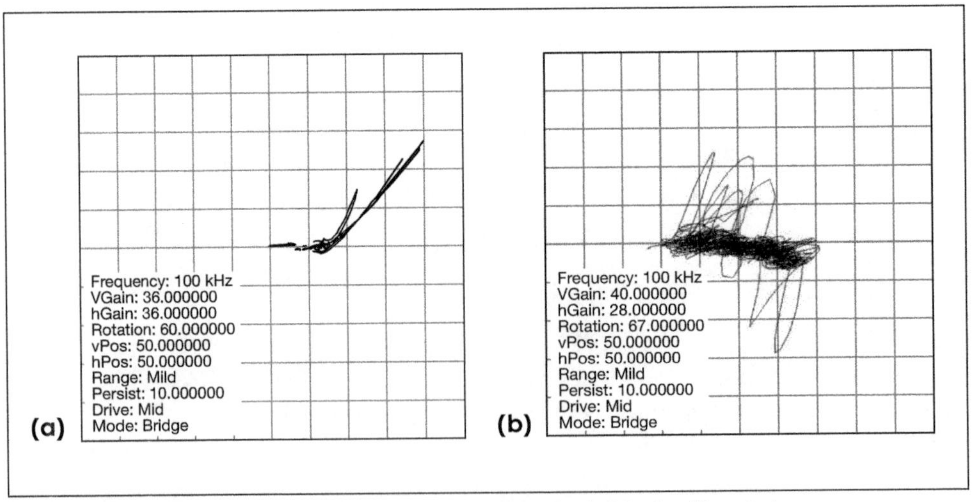

Figure 2. Electrical discharge machining notices: (a) in ferritic plate; (b) in complex weld crown (axial and transverse).

Weld scan probes are designed with two coils within the same probe body but with different coil orientations. The eddy currents generated by each is perpendicular to the other, as shown in Figure 3. One will detect axial indications only. The other will detect transverse weld cracks only. To maintain this orientation the technician must retain an established probe position or angle with respect to the weld geometry during the manual scanning process (Figure 4). Coverage is limited to just the area directly at the probe interface. Multiple scan passes are required to ensure complete coverage of the critical area.

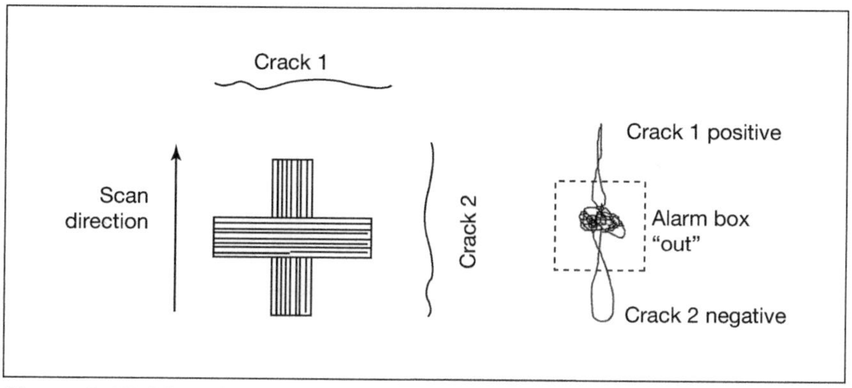

Figure 3. Weld probe coil sensitivity.

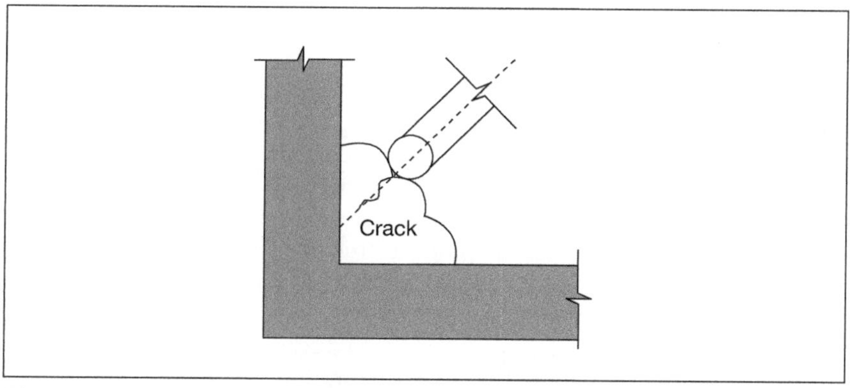

Figure 4. Weld probe orientation.

This type of coil arrangement is also very good for use in many other applications where geometry or localized material conditions might interfere with the coil's ability to detect small cracks.

Aerospace Applications

The aerospace industry uses many different types of eddy current inspections to detect a variety of discontinuities (Table 2). Most of the alloys used in constructing these complex systems are nonferromagnetic, with aluminum being the most prevalent.

Table 2. Aerospace discontinuities detectable by eddy current testing

Aerospace component	Discontinuities	Origin	Eddy current testing instruments
Airframe	Near-surface (skin) cracks	Flight stress	Small-diameter high-frequency (HF) pancake or pencil probes
	Subsurface and internal structural cracks	Flight stress	Larger-diameter low-frequency (LF) spot or driver/pickup probes
	Subsurface corrosion	Time, stress, and chemistry	Larger-diameter LF spot or driver/pickup probes
	Cracks at fasteners *Fasteners (in) *Fasteners (removed)	Flight stress	Pencil probes for skin layer *LF subsurface ring probes *HF rotating bolt-hole probes
Wheels	Cracks	Stress from repeated landings	Small-diameter HF pancake or pencil probes
	Signs of aging	Conductivity changes due to annealing processes generated by the heat of multiple landings	Larger-diameter LF spot or driver/pickup probes
			Special conductivity probes
Engines	Cracks	Flight stress	Small-diameter HF pancake or pencil probes
Welds	Cracks in ferritic and nonferritic alloys	Stress	Special probe coils optimized for weld inspection

Measurement of Metal and Coating Thickness

In the aerospace industry, numerous metal parts are coated or plated to obtain special surface properties such as corrosion resistance, wear resistance, or improved appearance. These coatings are prepared by various means, including electron deposition, hot dipping, cladding, and spraying. For controlling the thickness of the deposit and also for testing purposes, reliable and rapid measurements of coatings are necessary.

Four general types of coatings and base materials lend themselves to eddy current testing. The operating procedure used with the test instrument is determined by the specific combination of materials. These material combinations are classified as follows:

- Metal foil and sheet or metal coating on a nonconductive base material, such as metallic film on glass, ceramic, or plastics.
- Metal cladding with higher conductivity than the base metal, such as copper, zinc on steel, or pure aluminum on an aluminum alloy.
- Metal cladding with a lower conductivity than the base metal, such as nickel on aluminum.
- Nonconductive coating on a metallic base material, such as anodic film, paint on aluminum, or other organic coatings on metals.

Reference Standards for Thickness Testing

For eddy current thickness tests, at least three objects with known metal thicknesses are needed as reference standards to standardize the equipment. One reference standard represents the minimum acceptable thickness, the second represents the maximum acceptable thickness, and the third is from the middle of the range. All reference standards must have the same conductivity, permeability, substrate thickness, and basic geometry as the test objects.

Metal Thickness

Tests for measuring metal thickness are generally used on chemically milled sheet stock, thin-wall tubing, metal foil bonded to nonmetallic materials, and any objects that may experience thinning from corrosion.

Conductive Coating Thickness

Cladding thickness measurements are used in two types of situations where conductivities of two metal layers are very different. The first situation, in which the cladding is a better conductor than the base metal, pertains to copper, zinc, or cadmium coatings on steel base materials. The second situation, in which the cladding is the poorer conductor of the two metals, pertains to lead coatings on copper or to nickel coatings on aluminum.

Metal Spacing

There are times when a gap separates two metal sheets. The gap may be filled with a nonmetallic shim or may be purposely produced as a fixed dimension. If it is desired to measure the gap or spacing, an operating frequency must be chosen so that eddy currents will be generated in the second (subsurface) layer. The frequency chosen should produce eddy currents that penetrate both metals to a depth of 3δ, as shown in Figure 5.

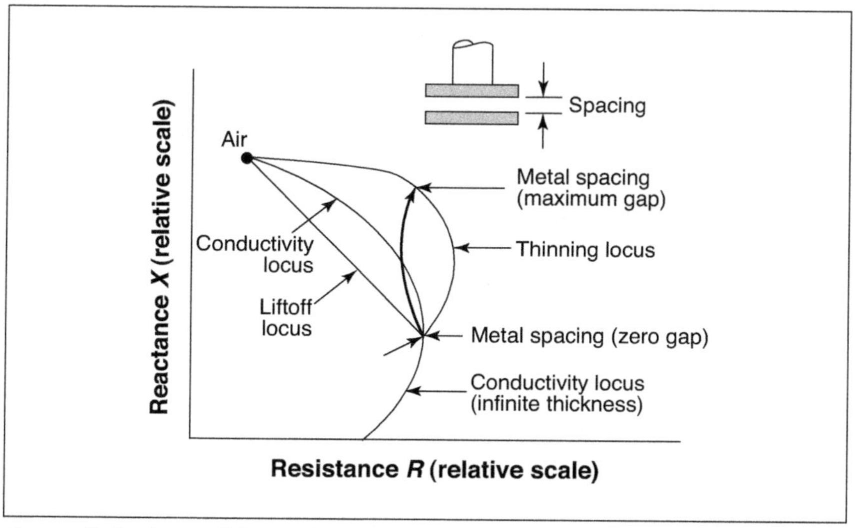

Figure 5. Probe coil impedance curve showing effect of metal spacing between two flat, parallel, metallic conducting sheets.

Gap or spacing measurements are best made with the probe located over the thin member; this reduces the problem with depth of penetration. If both members are equally thick, but one has a lower conductivity, the probe should be placed on the low-conductivity test object.

Tests of Metal Conductivity

Eddy current conductivity meters usually differ with respect to operating frequency, liftoff, temperature, sensitivity, and probe size compensation or means of presenting test results. Most conductivity meters can provide results as a percentage of International Annealed Copper Standard (%IACS). Other instruments can provide conductivity measurements in siemens per meter (S/m) or as customized displays of signal amplitude or phase angle (Figure 6). Digital test systems with numerical readouts are becoming much more common than the analog meters formerly used.

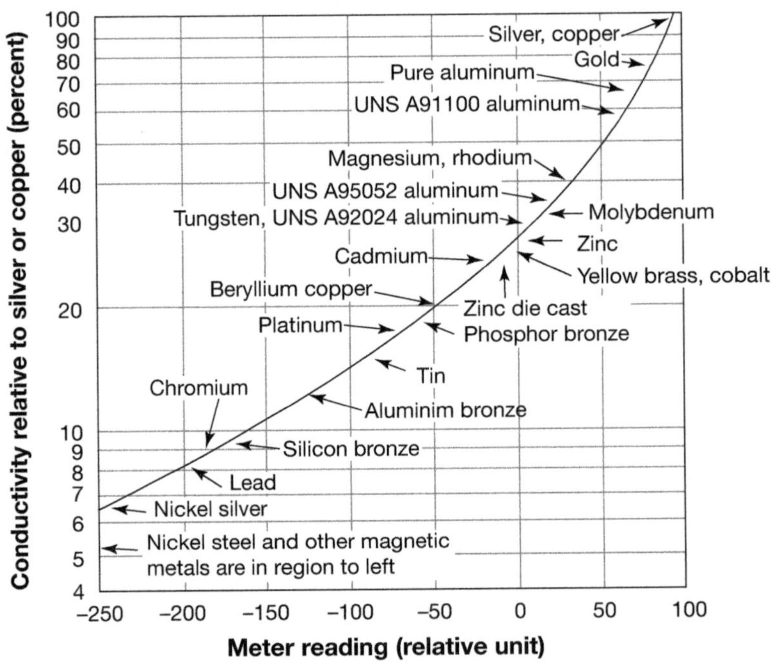

Figure 6. Relative conductivity of metals and alloys shown by eddy current meter readings. Unified Numbering System (UNS) is used for wrought aluminum alloys.

Testing of Bolt Holes

Small cracks in material located next to fastener holes may go undetected until the cracks grow to a size that allows detection at the surface not covered by the head of the fastener or nut. A crack of this magnitude and under high stresses may continue to grow to failure at a rapid rate. Eddy current testing is a reliable way to detect cracks in material adjacent to fastener holes. Eddy current tests are well known for detecting small fatigue cracks, particularly in aluminum fastener holes. Automated bolt-hole scanners provide reliable and repeatable test results. Finding small cracks at fastener holes sometimes requires removing the fastener and performing an eddy current check. Automatic scanning may be used to detect cracks as small as 0.005 in. (0.013 cm). Figure 7 illustrates an eddy current bolt-hole probe.

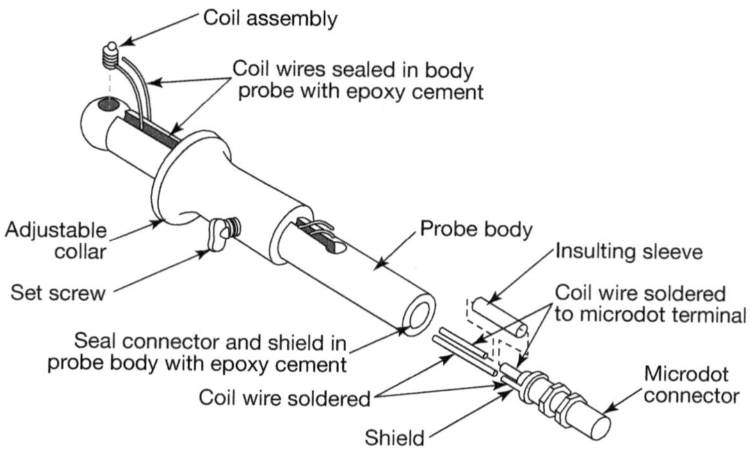

Figure 7. Eddy current bolt-hole probe.

Testing of Aircraft Structures

Eddy current testing is a primary nondestructive test method for detecting cracks and corrosion of aircraft structure and engine hardware during maintenance overhaul. These tests are performed in accordance with procedures developed by the manufacturer. Such tests can be used to detect fatigue cracks resulting from cyclic loading during flight, takeoff, or landing. They can also be used to detect stress-corrosion cracks to indicate the extent of corrosion damage or to identify portions of aircraft structures damaged by excessive heat.

Testing of Jet Engines

Eddy current testing is a well-established technique for the detection of surface discontinuities in conductive test objects. The US Air Force and other organizations have used automated eddy current testing to ensure that critical engine components are free of rejectable surface discontinuities.

Eddy current tests of military aircraft engines can be categorized as manufacturing or depot tests. When aircraft engine components are initially produced, eddy current tests of the manufactured objects are often required by specification to ensure that the new engine components are free of undesirable material properties or machining discontinuities. In-service parts similarly require depot tests at predetermined intervals to ensure that parts are free of small discontinuities in critical, predetermined zones. Manufacturing tests and depot tests are performed as preventive screening.

Surface Tests

Surface tests are also conducted using eddy current techniques to detect cracking in a variety of components that are difficult to test with other surface methods. Components typically tested include turbine rotors, turbine blades, socket welds, and stainless steel pipes susceptible to corrosion cracking of the outside surface. These applications typically require flat (pancake) coils or flexible array coils (Figure 8).

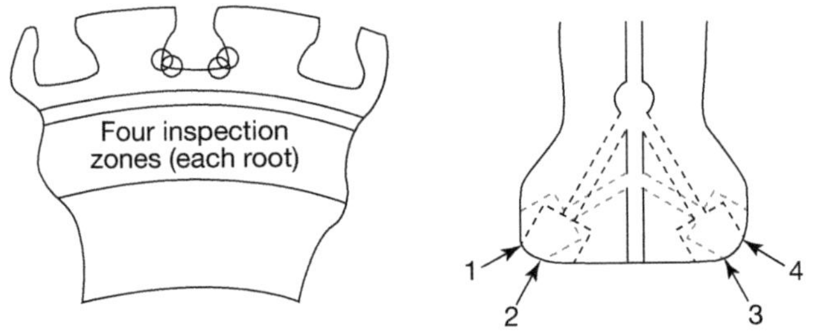

Figure 8. Probes and minimum coverage for turbine disk inspection.

Chemical and Petroleum Applications

The petrochemical market has many plants that cover very large areas. Within these facilities, a wide number of different operations may be performed. The classic example is to take crude oil and refine it to produce the many different fuels and lubricants required by the transportation sector. Many of these processes are performed in some type of heat exchanger. There may be hundreds of separate heat exchangers (HXs) in these operations (Figure 9). Typical discontinuities detected by eddy current testing are listed in Table 3.

Many of these HXs have arrays of tubes, ranging from just a few tubes to many hundreds or even thousands of tubes in each. These tubes will have one material form on the tube outside diameter and another material on the tube inside diameter. A material is being either heated up or cooled down for a specific reason. To ensure the overall quality of the product or plant safety, these tubes must keep these materials separated. If a penetration or leak occurs, it will result in the unit being brought offline until the issue can be resolved. This can take a matter of a few hours or several days. Lost production and revenue from these events can run into the hundreds of thousands of dollars.

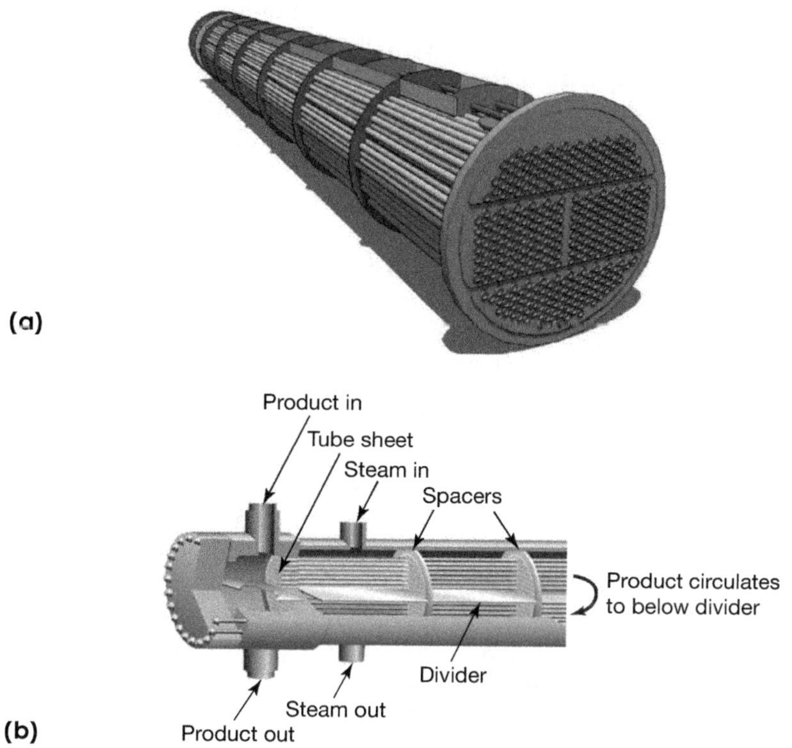

Figure 9. Heat exchanger: (a) tube bundle; (b) cutaway view showing process flow ports.

Table 3. Petrochemical discontinuities detectable with eddy current testing

Petrochemical component	Discontinuities	Origin	Eddy current testing instruments
Nonferritic tubing in a variety of different heat-exchanger designs	Multiple damage mechanisms in tube inside diameter (ID) and outside diameter (OD)	ID/OD chemistry	Bobbin coils with hardened outer cases
		Flow dynamics	
		Operational stresses	Frequency selection determined by tube alloy and wall thickness
		Mechanical vibration	
		Miscellaneous causes	
Welds	Cracks in ferritic and nonferritic alloys	Stress	Special probe coils optimized for weld inspection

Many petrochemical HXs are composed of ferritic (steel) tubing. Thus, eddy current will not work well on these units. These steel tubes would normally require the use of either remote field testing (RFT) or flux leakage testing. These two techniques are optimized for inspections of materials where eddy current testing may be limited.

The typical tool for nonferritic tubing is the classic bobbin probe. The frequency range for each probe is driven by its style and design. In thin-walled tubes of high-strength alloys (austenitic stainless steel or titanium), the frequency range will be higher than when the tubing is heavy walled or it has a higher conductivity (for example, with brass).

The following formula is used to determine what frequency range is required based on the tube wall thickness and the tubing alloy to be inspected:

(Eq. 1)
$$f_0 = \frac{K^2 \rho}{t^2}$$

where

f_0 is prime frequency (Hz)
K^2 is 10 (code-driven selection for ASME tubing examinations)
ρ is resisitivity (μΩ·cm)
t is wall thickness (in.)

Problem
Use the above formula to calculate the prime frequency for a tube made of admiralty brass ($\rho = 7.18$) with a wall thickness of 0.050 in. (1.27 mm).

Solution
Plugging in the numbers into the appropriate variables gives us the following calculation:

$$\begin{aligned} f_0 &= \frac{10 \times 7.18}{(0.050)^2} \\ &= \frac{70.18}{0.0025} \\ &= 28\ 072 \text{ Hz} \\ &\approx 28 \text{ kHz} \end{aligned}$$

Thus, the prime frequency for this type of tubing is approximately 28 kHz.

Electric Power Applications

Steam Generators

The most common use of eddy current testing in the power industry is tube testing in heat exchangers. Tubes in a nuclear steam generator are important for safety because they carry the primary cooling water. Degradation of steam generator tubing results from corrosion and mechanical mechanisms including intergranular corrosion, stress-corrosion cracking, thinning, impingement, and fatigue. Most tubes are fabricated from iron-nickel-chromium and nickel-chromium alloys that are nonferromagnetic.

Multifrequency bobbin coil techniques are used to test up to the full length of the tubes. It is recognized that the bobbin coil is not qualified for all tube locations, so other techniques such as rotating coil technology and array probes are also used.

Balance-of-Plant Heat Exchangers

In a nuclear power plant, HXs that do not carry reactor-cooling water from the containment vessel are commonly referred to as balance-of-plant heat exchangers. Most of them have counterparts in fossil fuel plants and are tested with multifrequency eddy current techniques in the differential and absolute modes using bobbin coil probes. Table 4 lists routinely inspected HXs in terms of tubing material and potential damage mechanisms.

Table 4. Heat exchanger materials and damage mechanisms

Heat exchanger	Tube material	Potential damage mechanisms
Feedwater heat exchanger	Carbon steel	Steam erosion
	Copper-nickel alloy	Pitting
	Stainless steel	Vibrational wear
		Condensate grooving
		Loose part damage
		Circumferential cracking
		Roll transition cracking
		Axial cracking
Generator stator cooling system	Alloy of 90% copper and 10% nickel	Inside surface pitting
Generator hydrogen cooler	Aluminum brass	Inside surface pitting and outside surface damage
Lubricating oil cooler	Admiralty brass	Inside surface pitting
Main condenser	Stainless steel-copper-nickel alloys; admiralty brass; titanium	Inside surface pitting and vibrational wear

Industrial Air Conditioning Chiller Applications

Cooling for large buildings and manufacturing processes is most commonly provided by centrifugal chillers (Figure 10). Condenser and evaporator tubing have fins on their outside diameters to increase the heat transfer surface, as shown in Figure 11. The heating, ventilation,

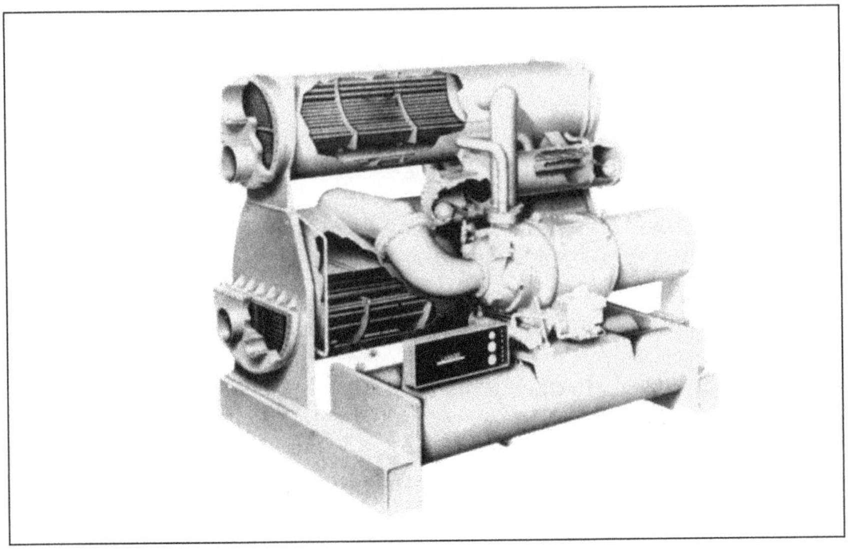

Figure 10. Centrifugal chiller.

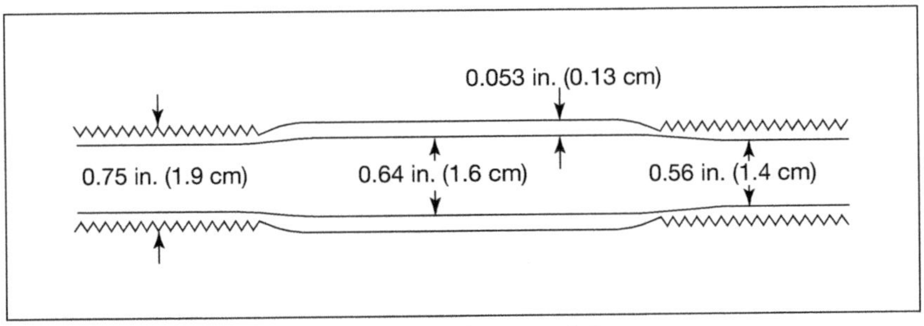

Figure 11. Cross section of heat exchanger tube.

and air conditioning (HVAC) market sector is heavily reliant on eddy current testing for inspection of the tubing in HXs (Table 5). Since the tubing is fairly heavy-walled and is made of highly conductive material (normally copper), a special family of bobbin probes has been developed for this arena.

The typical probes have a low-conductivity shielding element around the coils to promote longer life (Figure 12). They need to perform well in the low-frequency range (5 to 20 kHz). These probes are normally some combination of bobbin coils that allow for discontinuity detection in the finned transition zones where the normal geometry variations occur. They also have to be able to demonstrate discontinuity sensitivity to circumferentially oriented cracks. The second set of coils that are circumferentially sensitive are wound with their windings parallel to the tube's long axis or in an array or a cross-axis pattern.

The reference standards used in this market sector are often an issue. Effective eddy current screening of tube segments to be used as standards is essential (Figure 13). Copper is quite soft on a comparative scale of metals and is easy to damage during the machining operations. Poor standards make standardization and proper interpretation more difficult.

Table 5. HVAC discontinuities detectable with eddy current testing

Discontinuities	Origin	Eddy current testing instruments
Tube ID pitting	Chemistry	Low-frequency (LF) bobbin or special HVAC probes designed to compensate for tube-wall geometry variations
Tube OD wear	Vibration	LF bobbin or special HVAC probes
Cracks	Distorted geometry due to freeze bulges	LF bobbin or special HVAC probes

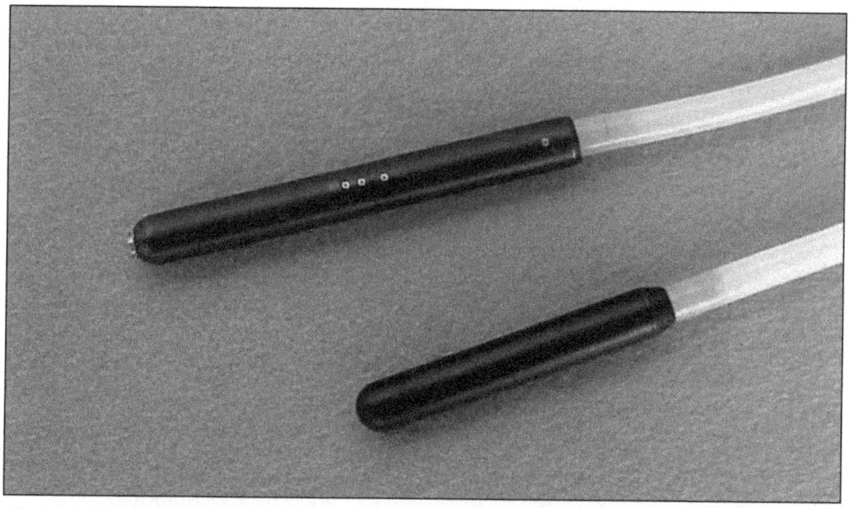

Figure 12. Eddy current probes for the inspection of HVAC tubing.

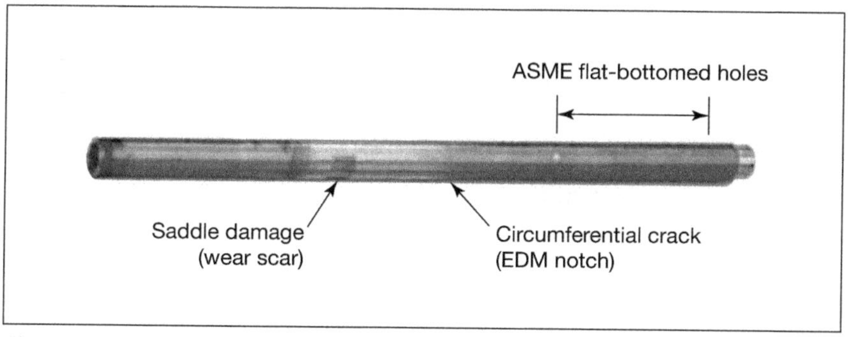

Figure 13. Reference standard for skip-finned copper tubing found in most commercial HVAC units.

Electromagnetic Testing in Primary Metals Industries

Electromagnetic testing techniques find their application in all stages of forming, shaping, and heat treating of metals and alloys, where the effectiveness of processing steps can be quickly evaluated. Materials damaged during processing can be detected and removed from production without incurring further processing costs. Thermal treatments such as annealing, normalizing, hardening, and other heat-treating processes can be directly monitored in many instances.

Material Sorting Related to Conductivity

Identifying or separating materials by composition or structure is referred to as *material sorting*. Impedance values are established on known reference material, and then the readings obtained from test references are compared with the reading obtained from the test object. Eddy current testing can also be used to identify and control heat-treatment conditions and evaluate fire damage to metallic structures.

Evaluation can also be accomplished by determining depths of case hardening of steels and some ferrous alloys. During age hardening of aluminum or titanium alloys, for example, the hardness and conductivity of the material change simultaneously so that the degree of hardening may be obtained by measuring the conductivity of the test object and comparing it with a standard of that material with a known hardness.

Testing of Hot-Rolled Bars

There are several requirements for an effective bar testing system:

- High discontinuity sensitivity.
- Ability to classify bar quantity.
- Full automatic operation.
- Rugged construction for use in mills.
- The ability to test bars as received without special preparation.

A rotating surface probe eddy current technique has been used for the testing of hot-rolled bars. With this technique, the equipment rotates an eddy current probe around an advancing bar. The probe is held a preset minimum distance from the bar surface. The equipment maintains the selected level of test sensitivity regardless of changes in discontinuity signal amplitude caused by varying surface spacing from probe to bar. It marks only those discontinuities that exceed a preselected length and depth. Bars with discontinuities are automatically separated from discontinuity-free bars.

Testing of Square Billets

Eddy current testing can automatically inspect 100% of the surface of steel billets without the need of an operator's judgment for interpreting test results. One such system can detect seams, evaluate their severity, and mark the location of those that exceed an acceptable depth. The key component is a scanning head assembly that keeps an eddy current probe in contact with, and tangent to, the billet surface at all locations around the periphery, including the corners. The machine is designed to test round-cornered, square billets as they are rolled. This integration with the manufacturing process is an important step in the development of integrated automatic testing and conditioning systems.

Testing of Hot Steel Rods and Wires

Surface discontinuity testing is essential in the quality assurance of iron and steel products. In many mills, quality control of hot-rolled rods is provided through eddy current and flux leakage testing carried out after the rolling, shearing, and cooling process. Wires and bars are usually coiled immediately after they are hot rolled. When discontinuity detection is performed on cold products, they must be uncoiled for testing. Eddy current testing using an encircling coil has been applied to hot rolling of bars. Steel mills use encircling-coil eddy current systems to test hot wires. Generally, encircling coils are in the differential mode coil arrangement. This testing system can detect short discontinuities, such as scabs and roll marks. A diagram of a single-strand steel rod control system is shown in Figure 14.

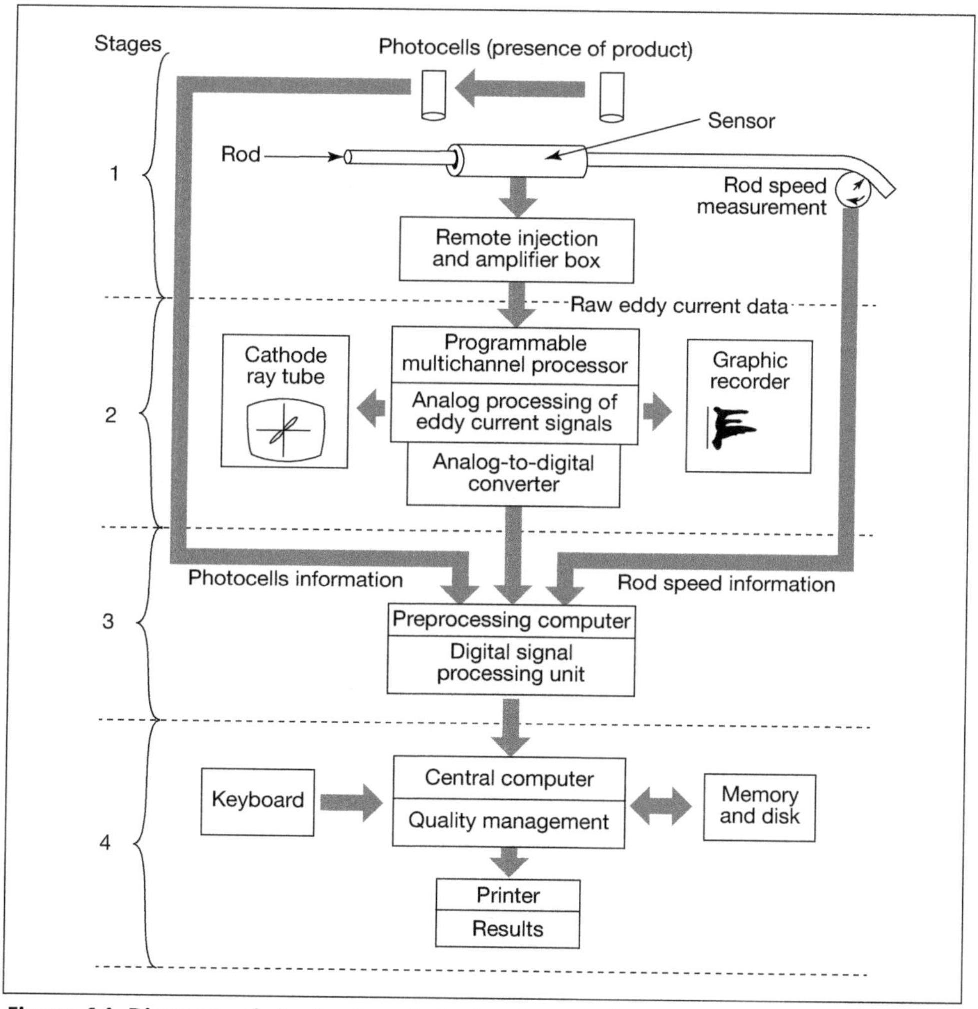

Figure 14. Diagram of single-strand steel rod control system showing four stages: (1) signal acquisition; (2) data display; (3) data sorting; (4) reporting.

12

Alternating Current Field Measurement Testing

History

In the 1980s, there were moves to develop nondestructive testing (NDT) techniques for detecting and sizing fatigue cracks underwater in welded offshore structures. Oil companies working in the North Sea were looking for a technique better suited to estimating the depths of fatigue cracks found underwater by magnetic particle testing (MT). Alternating current potential drop (ACPD) was one technique used for this purpose at the time, but it was very difficult and slow to use underwater because of the need to maintain very good electrical contact between the voltage probe and the surface.

A noncontact equivalent to ACPD was required. The mechanical engineering department at University College London (UCL), United Kingdom, had many years' experience in both theoretical and practical developments in ACPD, so a group of UK oil companies approached it to develop the new technique, which became known as alternating current field measurement. (*Note:* The acronym ACFM is a registered trademark of TSC Inspection Systems.)

Alternating current field measurement uses induced rather than injected currents but tries to maintain the same uniform-strength, unidirectional currents present for ACPD. This allowed UCL to take existing theoretical models of the effect of semielliptical surface-breaking discontinuities on the current flow and to extend them to predict the effect on the associated magnetic fields above the surface. By measuring components of these magnetic fields and comparing the results with the theoretical predictions, alternating current field measurement is able to determine the length and depth of a discontinuity.

The theoretical model and early experimental results were published in 1988, and a new inspection technique was born. Over the following years, alternating current field measurement has expanded into a useful addition to the NDT toolbox, particularly for inspecting painted or coated welded steel structures.

Principle of Operation

The alternating current field measurement technique involves inducing a locally uniform current into a test object and measuring the magnetic flux density above the test object surface. The presence of a surface-breaking discontinuity perturbs the induced current and hence the magnetic flux density.

Two components of this magnetic field are measured: one provides information about the depth and aspect ratio of the discontinuities; the other provides information on the positions of the ends of each discontinuity. The two signals are then used together to confirm the presence of a discontinuity, which, along with a sizing algorithm, predict the surface length and depth of the discontinuity.

The technique uses an instrument and a handheld probe containing a uniform field induction system and two magnetic field sensors, which measure two orthogonal directions of the magnetic field. Software is used to control the instrument and to display, analyze, and store the resultant data. It should be noted that, although the technique uses magnetic field measurements to detect and size discontinuities, its application is not restricted to ferromagnetic metals.

Advantages and Limitations

The alternating current field measurement technique uses a uniform input field to allow comparison of signal intensities with theoretical predictions. A uniform field has advantages and limitations compared with conventional eddy current testing.

Advantages

The main advantages include the following:

- The ability to test through coatings up to 0.2 in. (5 mm) thick.
- The ability to obtain depth information on cracks up to 1 in. (25 mm) deep.
- Easy testing at material boundaries, such as welds.
- No requirement for onsite standardization.

Limitations

The main limitations include the following:

- Lower sensitivity to smaller discontinuities. This reduction in sensitivity is of little consequence on welded or rough surfaces where sensitivity would be reduced anyway. On smooth, clean surfaces, however, alternating current field measurement is generally less sensitive to short or shallow discontinuities than conventional eddy current testing.
- Geometry changes such as plate edges and corners can produce signals that may confuse the technician, even though these signals do not have the same form as a signal from a discontinuity.

- Discontinuity's orientation relative to probe. The signals obtained from a discontinuity depend on the orientation of the discontinuity. The uniform field theoretical model suggests that no signal is produced when a probe scans across a transverse discontinuity, because the current flow is then parallel to the discontinuity and would not be perturbed.

Alternating Current Field Measurement Specific Theory

As alternating current field measurement is an electromagnetic technique, the general theories of magnetic fields, electrical circuits, and induction, discussed in previous chapters, apply here also. This section is concerned with the specific theory of alternating current field measurement operation and particularly where it differs from conventional eddy current testing.

The required locally uniform magnetic field is induced using one or more horizontal axis solenoids, with or without a core. The inducer is energized using a *constant-current sinusoid* of a selected frequency, typically 5 kHz. The constant current driver ensures that the magnetic flux generated by the solenoid remains constant regardless of impedance changes in the solenoid due to changes in material property or liftoff above the surface.

Figure 1 illustrates the magnetic field generated around a horizontal solenoid when positioned above a metal surface. As with other induced electromagnetic techniques, the magnetic field is confined to a layer in the surface of the material defined by the material permeability, conductivity, and test frequency. The so-called *skin depth*, δ, which corresponds to the depth where the amplitude of the field density is approximately 37% of that on the surface, is given by:

(Eq. 1)
$$\delta = \frac{1}{\sqrt{\sigma \times \mu_0 \times \mu_r \times f \times \pi}}$$

where

σ = standard depth of penetration, in meters
f = frequency, in hertz (Hz)
μ_0 = magnetic permeability, in henries per meter (H/m)
μ_r = relative permeability, in henries per meter (H/m)
π = 3.1416

The alternating magnetic field in the skin of the metal induces a current to flow on the surface of the metal at right angles to the magnetic field and in the plane of the metal surface. Immediately below the inducer, the magnetic field is essentially linear and of consistent density.

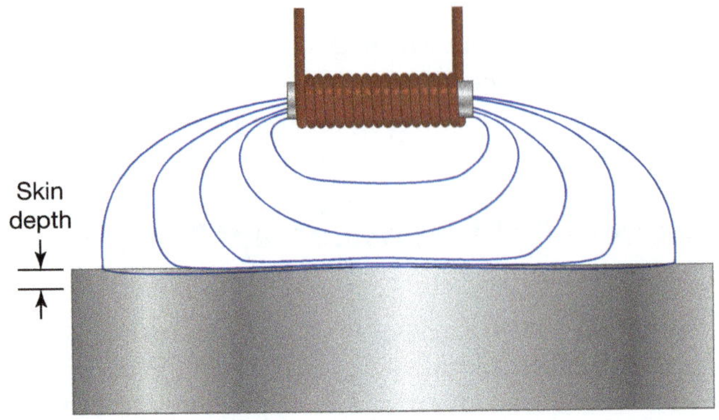

Figure 1. Magnetic field generated by horizontally oriented solenoid.

Relation of Magnetic and Electric Fields

If we consider a uniform AC sheet in the surface of a material, as in Figure 2a, this will itself produce a magnetic field above the surface at right angles to the current direction.

With reference to the coordinate system in the figure, where the X and Y directions are along the surface of the material and the Z direction is normal (perpendicular) to the surface, a current flowing in the Y direction will produce a magnetic field (symbol B) in the X direction. We term this B_x.

The magnitude of the B field is proportional to the current density in the electric field: the higher the current density (the closer the flow lines are together), the higher the magnetic field.

In Figure 2a, the B_y and B_z components of the magnetic field are both zero and the B_x level depends on the magnitude of the current.

If the current flow lines are parallel, then there is no component produced in Z, that is, B_z is zero. A resultant of the magnetic field in the Z direction is produced by curvature of the current or bending of the

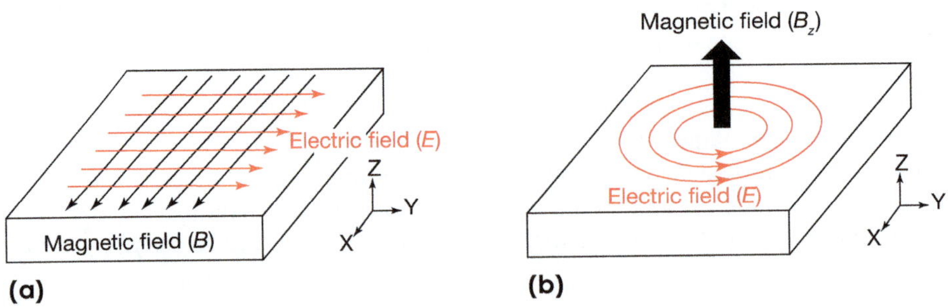

Figure 2. Magnetic field: (a) produced by current in plate; (b) due to current curvature.

flow lines, as shown in Figure 2b. B_z will be positive if the current curves in one direction and negative if it curves in the other direction.

The important principles to remember are that if a current is flowing in a surface (called the X-Y plane), then:

- The magnetic flux density in the X direction is proportional to current in the Y direction.
- The magnetic flux density in the Y direction is proportional to current in the X direction.
- The magnetic flux density in the Z direction (out of the X-Y plane) is proportional to the curvature of the current in the X-Y plane.

Fields Around a Discontinuity

If a surface-breaking linear discontinuity is introduced in the material surface such that it is oriented in the X direction and cuts across the line of the electric field, shown in Figure 2a, the electric currents will not be able to flow across the high resistance of the discontinuity air gap. The presence of the discontinuity will distort the electric field underneath and around the ends of the discontinuity.

Some of the current is forced away from the surface, as it flows underneath the discontinuity. This reduces the strength of the magnetic field in the middle of the discontinuity, shown in blue in Figure 3. Some of the current flows around the ends of the discontinuity—that is, the increased current density strengthens the magnetic field at the ends—shown in red.

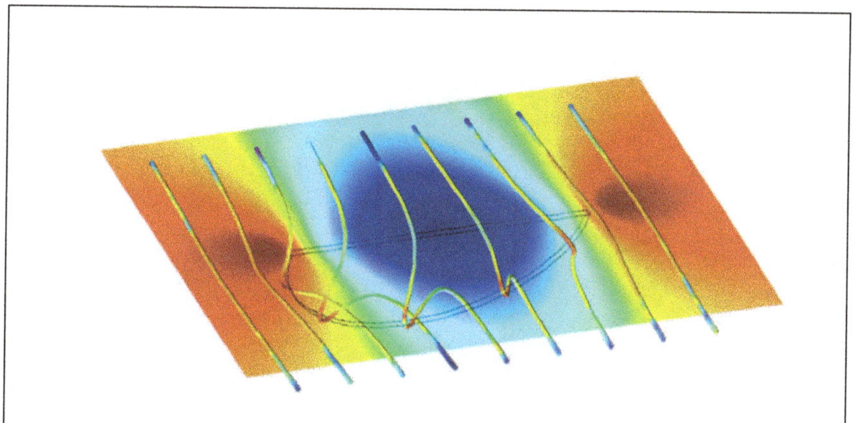

Figure 3. Disturbed electric field around a linear discontinuity.

The horizontal component of the magnetic field is shown in 3D in Figure 4a. The blue dip in the curve relates to the depth of the discontinuity. Magnetic sensor coils in the probe measure the distribution along the length of the discontinuity and produce the B_x curve in the alternating current field measurement software, as shown in Figure 4b.

As the current is forced to flow around the ends of the discontinuity, it has a slight rotation to its motion. This rotation around the ends of the

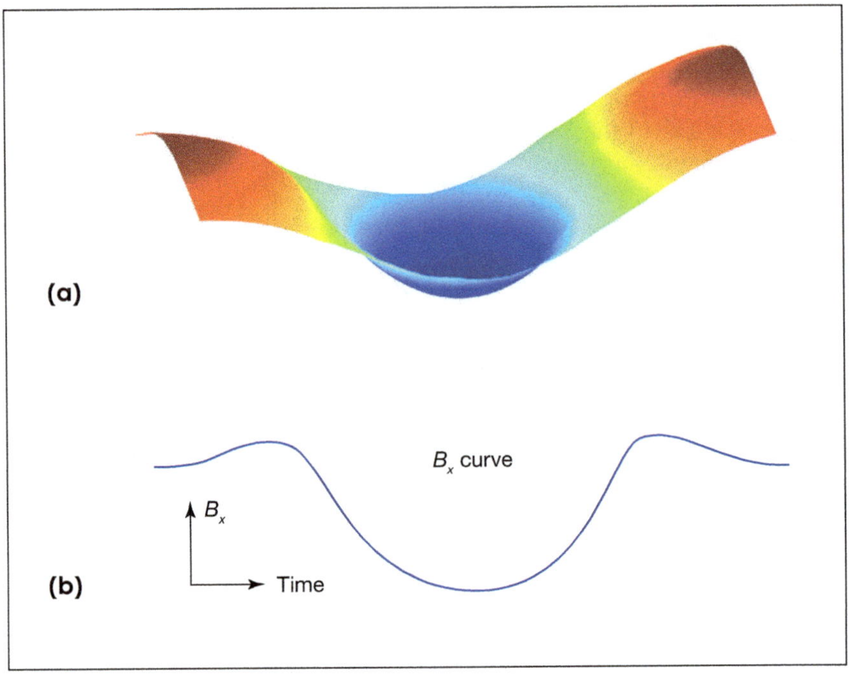

Figure 4. Generation of B_x curve: (a) 3D representation; (b) graph.

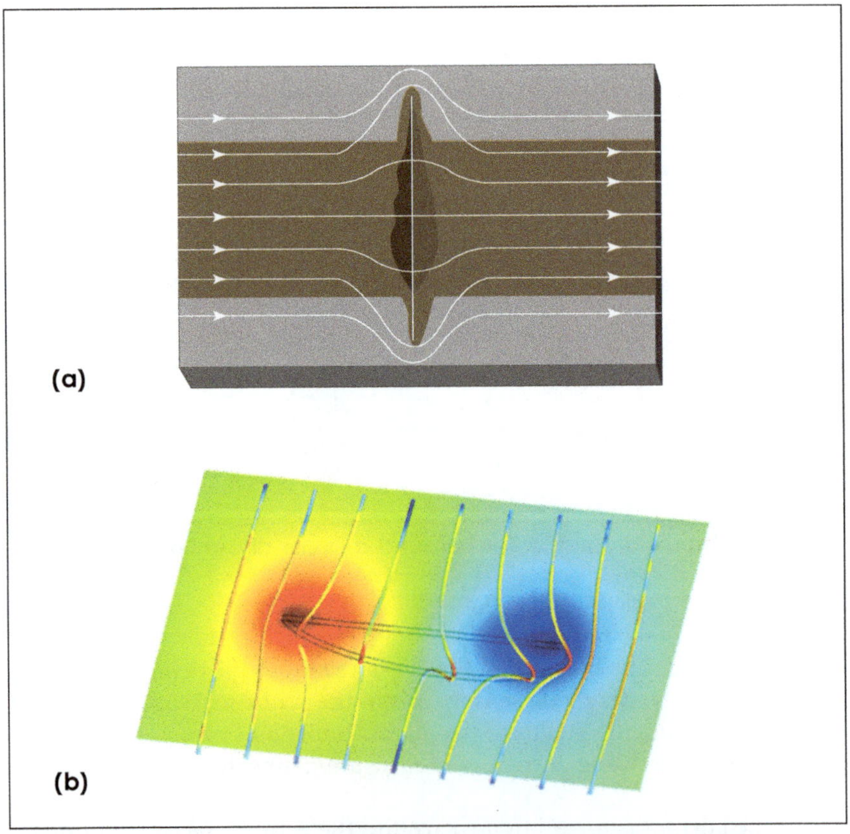

Figure 5. Rotation of current density around ends of discontinuity: (a) schematic; (b) 3D representation.

crack produces a magnetic field perpendicular to the surface, as shown in Figure 5a.

As the current rotates clockwise at one end of the crack, it produces a negative vertical field, blue in Figure 5b; rotating counterclockwise at the other end produces a positive vertical field, red in Figure 5b.

The vertical component of the magnetic field is shown in 3D in Figure 6a. The peak and trough mark the beginning and end of the discontinuity. This produces the B_z curve in the alternating current field measurement software, as shown in Figure 6b.

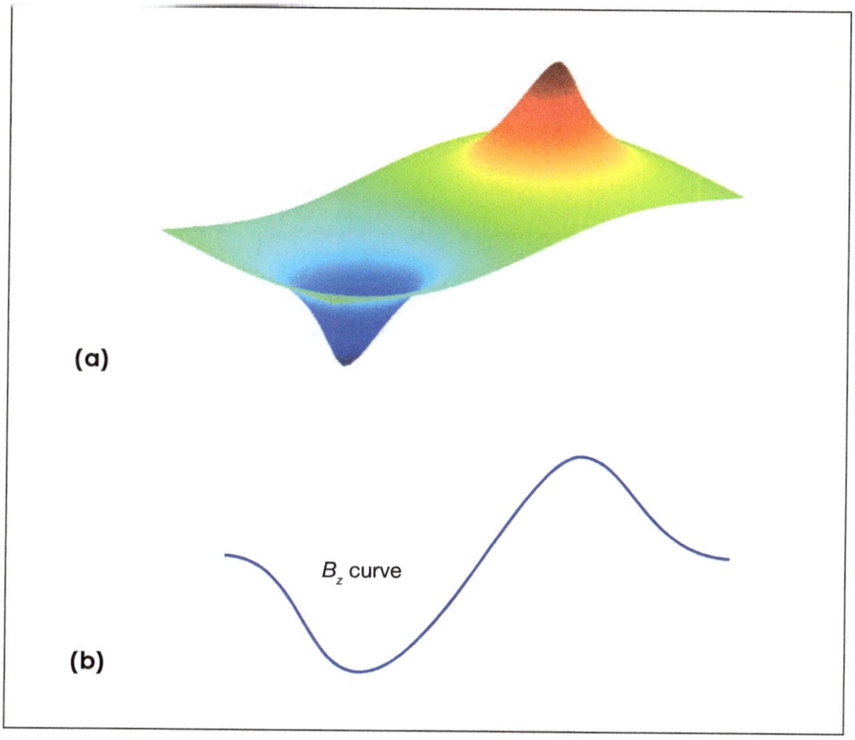

Figure 6. Generation of B_Z curve: (a) 3D representation; (b) graph.

The Butterfly Plot

A third plot is normally shown on the alternating current field measurement software screen, called the *butterfly plot*, shown in Figure 7. While this initially looks like an eddy current phase plot, it is actually created from the B_x and B_z sensor data plotted against each other. The B_x amplitude is plotted on the vertical axis and the B_z amplitude is plotted on the horizontal axis (for historical reasons the positive direction of the B_z axis is to the left rather than to the right as in a conventional cartesian graph).

When a discontinuity indication is produced by the B_x and B_z sensors, the effect on the butterfly plot is to produce a downward loop. The direction of this loop coincides with the direction mark displayed on the probe: if the probe is moved in one direction, then the butterfly produces a clockwise loop; if the probe is moved in the opposite

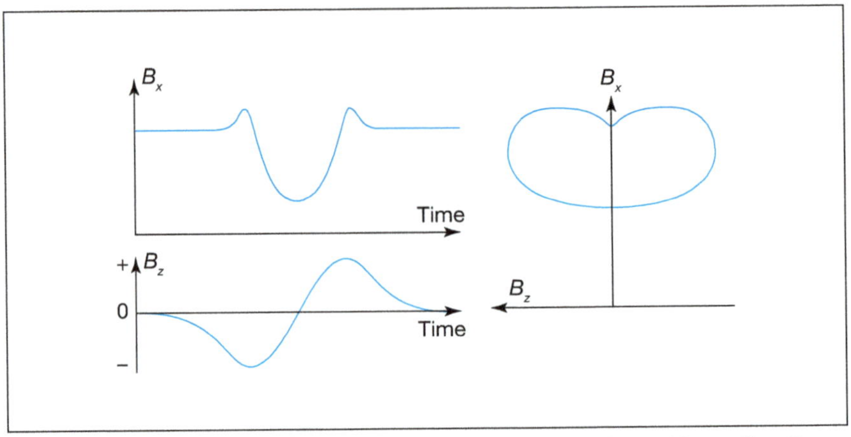

Figure 7. The butterfly plot from a discontinuity produced from the B_X and B_Z traces.

direction, the loop is created in a counterclockwise ("anti-clockwise" in UK English) direction.

The resultant display effectively removes the time element from the scan since only the relative B_x and B_z amplitudes are plotted, which is more closely associated with a particular location on the sample.

Alternating Current Field Measurement Equipment

Basic Alternating Current Field Measurement Probe

A basic alternating current field measurement probe contains the following three elements at minimum:

- A magnetic field inducer.
- A magnetic sensor oriented to measure flux density in the X direction.
- A magnetic sensor oriented to measure flux density in the Z direction (normal to the surface).

Most alternating current field measurement probes also contain active electronics that amplify and buffer the sensor signal as well as store the probe's serial number, which can be read by the controlling instrument.

The field inducer typically consists of a solenoid housed in the body of the probe and oriented parallel with the test surface (Figure 8). The solenoid may have a steel, ferrite, or air core, and the length/height (above the surface) ratio largely dictates the uniformity of the magnetic field at the nose of the probe.

The two sensors are typically air-cored coils, which are mounted in the nose of the probe and oriented such that one coil is parallel to the surface with its axis in the expected discontinuity direction (B_x) and the other is oriented vertically (B_z) (Figure 9).

As discussed, the B_x sensor measures the reduction in flux density at the middle of a discontinuity, which is primarily related to the discontinuity depth, and the B_z sensor measures the rotation in current

Figure 8. Field inducer in pencil probe.

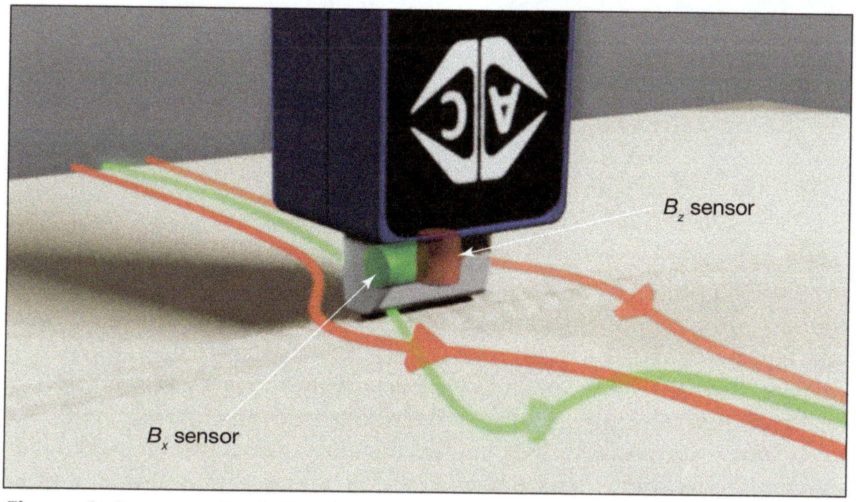

Figure 9. Sensor orientation in pencil probe nose.

that occurs at the ends of a discontinuity and so provides information on discontinuity length.

Alternating Current Field Measurement Array Probes

Basic alternating current field measurement probes, as described on the previous page, inspect a narrow strip between 0.4 and 0.6 in. (10 and 15 mm) wide, centered on the probe sensors. In some situations, it is desirable to cover wider areas with a probe. Alternating current field measurement array probes contain multiple sensor pairs arranged adjacent to each other, which allow a larger area to be covered when the probe is scanned on the component. The sensor array may be arranged with a flat profile, a custom-shaped profile, or as an adaptable, conforming profile (Figure 10).

Some array probes also feature an additional field inducer (Y field) arranged normal (perpendicular) to the primary field inducer (X field) together with an additional sensor, the B_y sensor. The fields are switched electronically as the sensor channels are multiplexed to produce two sets of data: one set taken using the X field and the other using the Y field. These probes are capable of inspecting in orthogonal directions at the same time without the need to rotate the probe.

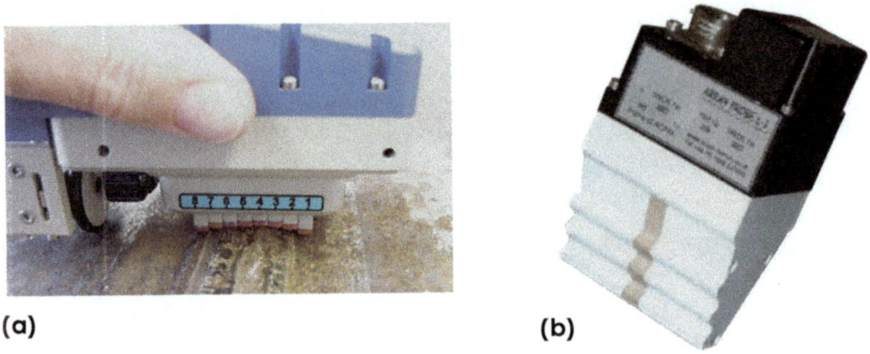

(a) (b)

Figure 10. Arrays: (a) conforming profile array; (b) custom-shaped array.

The data produced by array probes are fundamentally the same as for basic probes but with additional traces for each pair of sensors. The data can be viewed as a traditional trace plot or as a color C-scan type of display, where red peaks and blue troughs are shown against a green background level. Figure 11 shows the results from an eight-element array probe scanned over four longitudinal linear discontinuities and one transverse oriented discontinuity.

Alternating Current Field Measurement Instrumentation
Alternating current field measurement instruments are broadly categorized by their use above water or underwater. Two common instruments are shown in Figure 12.

Discontinuity Types

Fatigue Cracks
Alternating current field measurement was designed for the detection and sizing of fatigue cracks. Fatigue cracks are generally surface-breaking discontinuities that tend to grow at defined stress concentrations and are well suited for the linear scanning path of alternating current field measurement probes. Fatigue cracks also tend to grow in a semielliptical shape and at right angles to the surface.

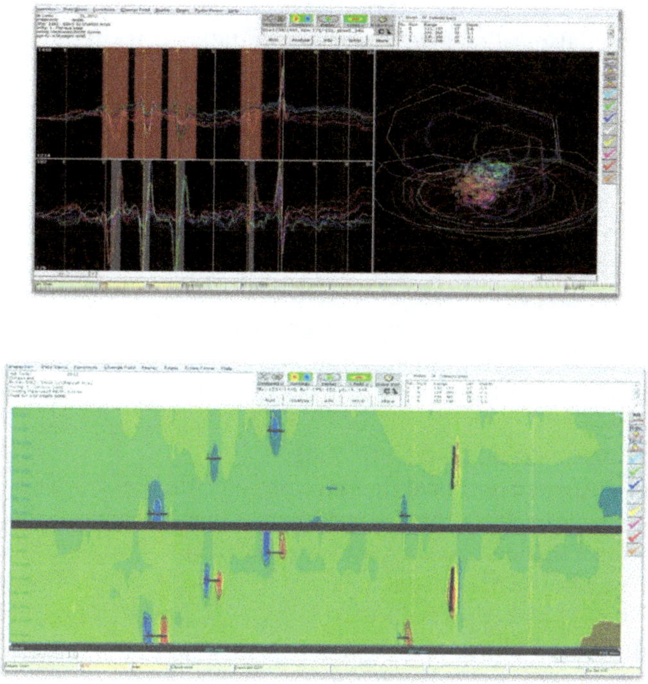

Figure 11. Different data displays using alternating current field measurement arrays.

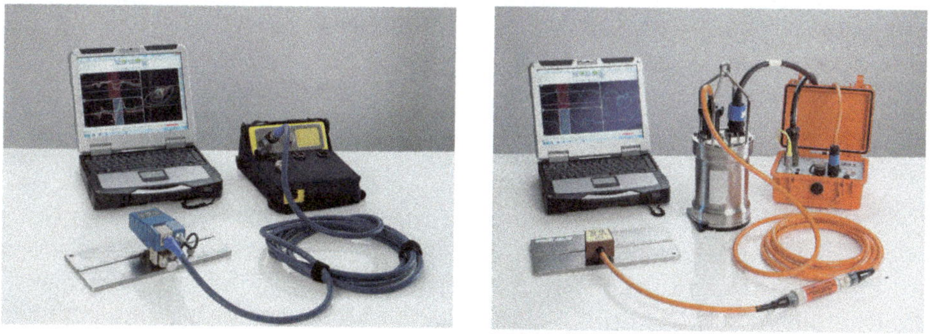

(a)

(b)

Figure 12. Alternating current field measurement instrumentation: (a) above water; (b) underwater.

Fatigue Cracks in Rail Heads

Non-semielliptical discontinuities occur also in railroad rails. *Head checking* (also called *gauge corner cracking*) is cracking that initially grows into the top surface of a rail at a highly inclined angle. As these cracks grow below a certain depth, they turn to a steeper angle but also start to grow sideways, so their length is greater beneath the surface. Alternating current field measurement can be used to detect this type of cracking.

Stress-Corrosion Cracking

Stress-corrosion cracking can take the form of a series of parallel cracks acting as a colony. The detection of stress-corrosion cracking by alternating current field measurement is reliable, and depth values obtained by treating isolated clusters as single discontinuities agree reasonably well with the typical discontinuity depth.

Hydrogen-Induced Cracking

Alternating current field measurement techniques have also been used to detect sulfide stress concentration cracking, hydrogen-induced cracking, hydrogen sulfide cracking, and stress-oriented hydrogen-induced cracking in the base metal adjacent to the heat-affected zone.

Corrosion Pitting

The unidirectional currents used in alternating current field measurement are most strongly perturbed by planar discontinuities. However, surface-corrosion pitting perturbs current flow to some extent and can also be detected. The degree of current perturbation is much lower than for a crack of the same depth and length, so on an initial scan, a corrosion pit looks like a shallow crack. However, the distinguishing feature of a pit is that, unlike a crack, it produces the same signal regardless of the orientation of the interrogating current.

Signal Interpretation

Transverse Discontinuities

Alternating current field measurement uses a unidirectional current, so responses vary with discontinuity orientation. A discontinuity that is transverse to the probe scan direction will be parallel to the current, and therefore no current perturbation will occur. This means that normally no signal is seen. This is always true for nonferritic metals, but for ferritic metals, it is possible to get a flux leakage effect, whereby the magnetic field (at right angles to the current) crosses the discontinuity, essentially leaking out into the air above the discontinuity. This effect depends on the opening (width) of the discontinuity, so it is more likely to be seen on slots or deep cracks. The B_x signal goes up as the probe scans over the discontinuity, while the B_z signal exhibits a sharp peak/trough pair. This combination produces an upward-going loop in the butterfly plot (Figure 13). It is not possible to size a discontinuity from this signal. To do this, the probe must be turned and scanned along the line of the discontinuity.

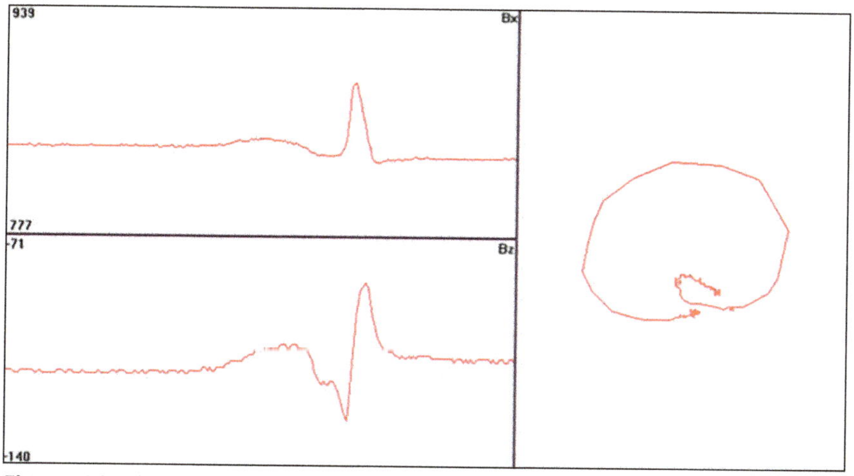

Figure 13. Signal from a transverse discontinuity in ferrous steel.

Edge Effects

The large-scale nature of the induced current in alternating current field measurement means that the current is distorted as a probe approaches a free plate edge. This gives rise to a large change in the signal, particularly for B_x (Figure 14). The distance from the edge at which this occurs depends on the probe design, being larger for the larger weld probe.

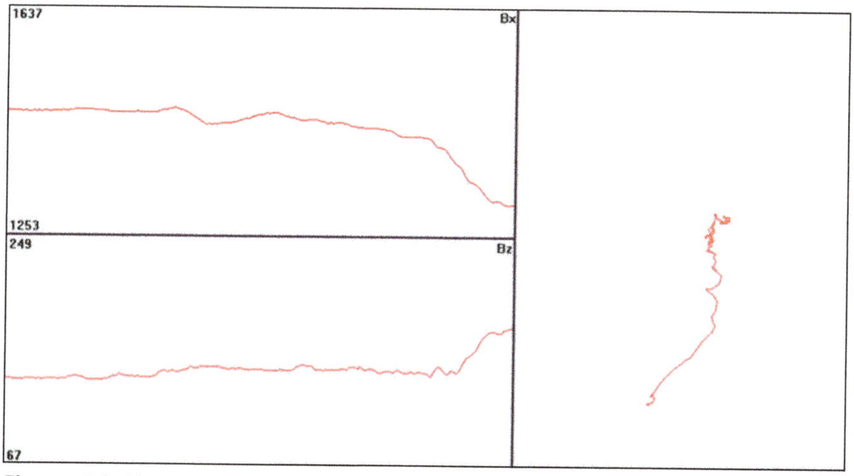

Figure 14. Signal from a probe scanned toward a plate edge.

A discontinuity indication close to an edge will still create a butterfly shape, but the background values, marked A and B on Figure 15, will be at different levels. In this case, the two ends of the butterfly plot will not meet. This makes these discontinuities harder to identify.

Neither background level (A or B) is correct for sizing. To place the B_x background level correctly, the operator must assess where the background level would be if there were no crack. This is shown by the dashed arc and is often taken as the average of B and A.

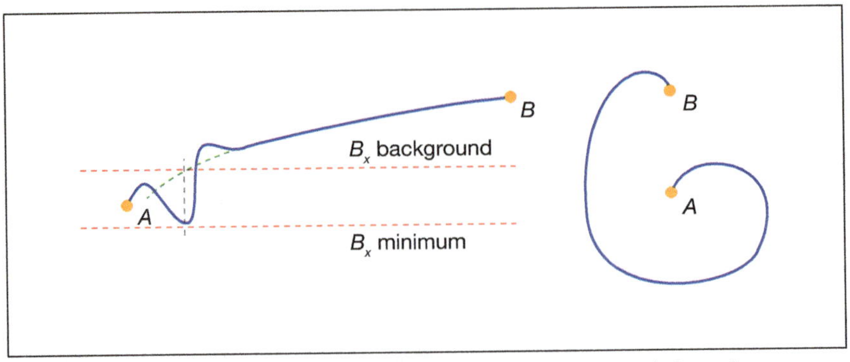

Figure 15. Example signal from a discontinuity near a plate edge.

Cracking into Parent Plate

In some geometries, especially tubular intersections, long discontinuities that start growing along a weld toe can deviate into the parent plate. Depending on the orientation of the discontinuity and probe scan direction, this will be shown as a sudden drop in B_x with no corresponding B_z peak or trough, marking a discontinuity start. Alternatively, the start of a discontinuity signal may be seen, but then the B_x rises up to the background level without an accompanying B_z end signal (Figure 16). It is also possible that a discontinuity may deviate both into the parent plate and into the weld. In this situation, no B_z end signals would be seen on the toe scan, and the B_x dip would be the only initial indication.

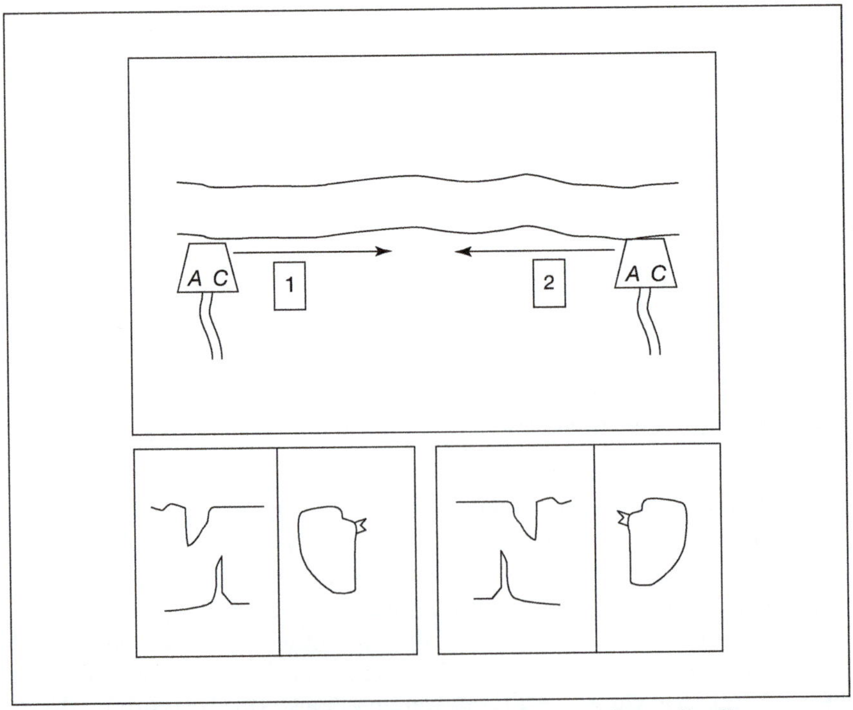

Figure 16. Examples of indications produced by a discontinuity deviating into parent plate.

If a deviating crack is suspected, scans should be made on the parent plate (or weld, as appropriate), along the expected discontinuity line, to find the discontinuity's ends. Where the expected discontinuity line is not known, then the probe can be scanned back and forth parallel to the weld but gradually moving away from it. The discontinuity line can be tracked by noting the positions of the B_x dips as the probe crosses the discontinuity. The end of the discontinuity can be found by watching for the B_z peak or trough, as appropriate, and maneuvering the probe until the maximum signal is seen. Note that if the discontinuity turns transverse to the weld line, then the B_x signal may rise as the probe crosses the crack in ferrous materials.

Seam Welds

Seam welds running across the line of scanning also produce strong signals in B_x and B_z, which can sometimes be confused with a discontinuity signal. Depending on the weld material of the seam, it may produce an upward or downward butterfly loop (Figure 17).

As for grinding marks, if a discontinuity is suspected at or near a seam weld, further scans should be made away from but parallel to the suspected discontinuity line to confirm the discontinuity. The signal

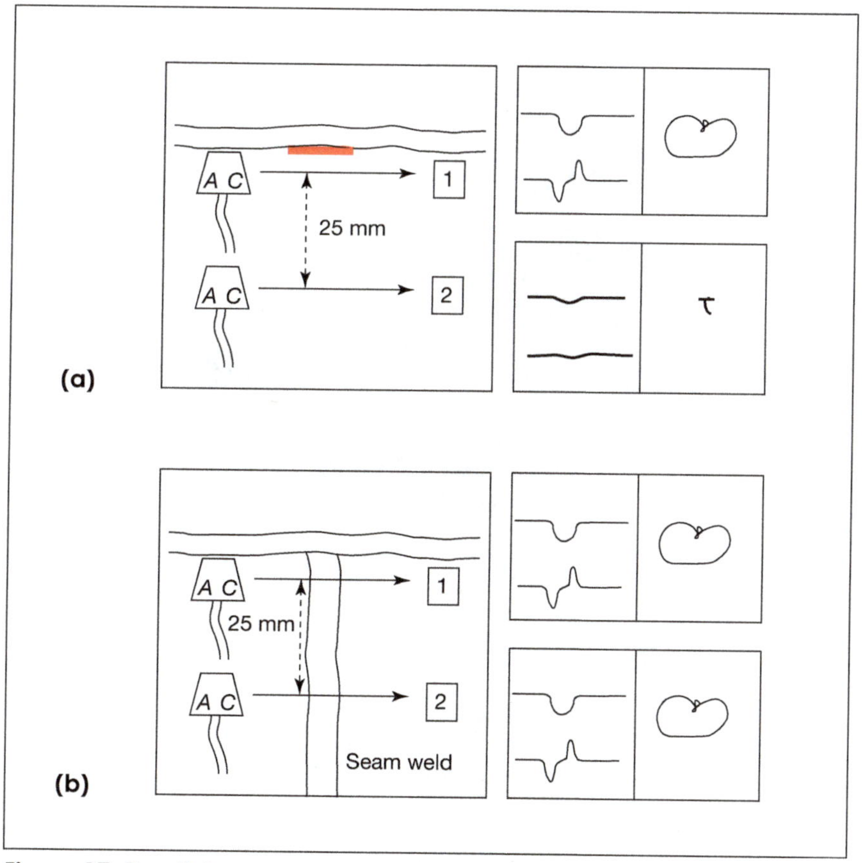

Figure 17. Parallel scan responses: (a) with crack present at toe; (b) where a seam weld is present.

from the seam weld will persist at roughly the same amplitude, whereas a discontinuity signal will exhibit significantly reduced amplitude. Note that seam welds are often ground flush and can be difficult to see, particularly under water. For this reason, scans parallel to the weld should be carried out for any suspected discontinuity signal if seam welds might be present.

Typical Applications

Manual Weld Inspection

Alternating current field measurement is used for the inspection of welded connections for surface-breaking discontinuities in a wide range of industries.

Some of the work involves the use of two-person teams, including rope access specialists. With two-person operations, alternating current field measurement allows the probe technician to be remote from the inspector (Figure 18). The use of rope access avoids the need for

(a)

(b)

Figure 18. Alternating current field measurement performed: (a) on coated pipework; (b) via rope access.

scaffolding. Alternating current field measurement also avoids the need for paint removal and reapplication. The butterfly plot removes the effect of nonuniform probe movement to allow reliable use of non-inspectors as access providers. In this way, it is not necessary to have skilled inspectors who are also skilled climbers or divers (Figure 19).

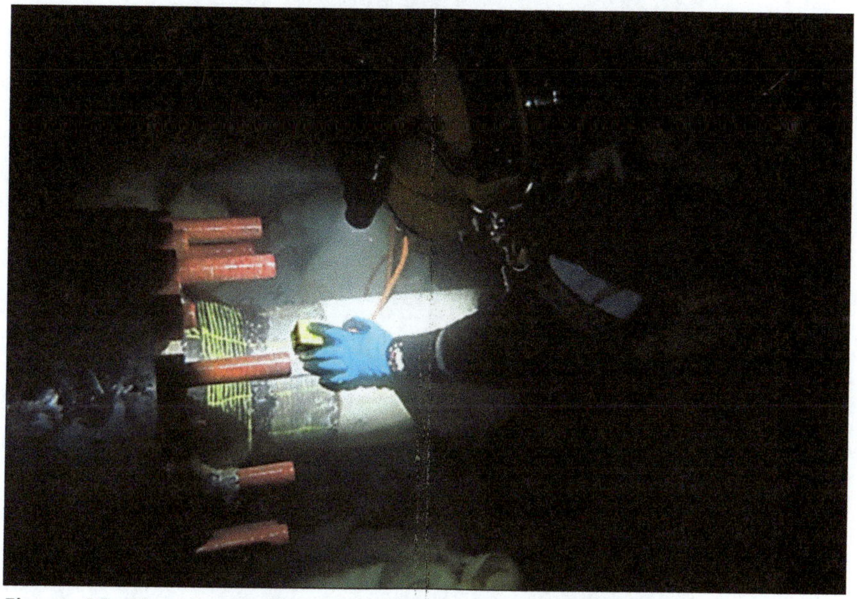

Figure 19. Diver inspecting a subsea flange-pipe weld.

Thread Inspection

Alternating current field measurement can be used to inspect threaded connections, particularly those used in onshore and offshore drilling. Compared to conventional inspection methods (MT and PT), benefits include:

- Reduced cleaning requirement compared to MT and PT.
- Ability to work on ferromagnetic and nonferromagnetic components.
- Can operate under wide lighting conditions.
- Provides information of crack depth in addition to location and length.
- Easy transfer of data in electronic form for review offsite, if required.

Limitations and disadvantages include:

- Multiple probes are required for various thread forms.
- Inspection of the stress relief groove requires additional scans.

Thread array probes are available configured to a particular thread pitch, typically 3.5, 4, or 5 threads per inch (tpi). Adapters are fitted to either side of the probe to position the thread sensors correctly for the thread size (pin or box) to be inspected (Figure 20). Integral stainless steel scrapers engage in the thread and closely follow the thread profile, as displayed in Figure 21a. The scrapers remove dirt and contaminants from the thread and prevent wear to the thread sensors, which are held

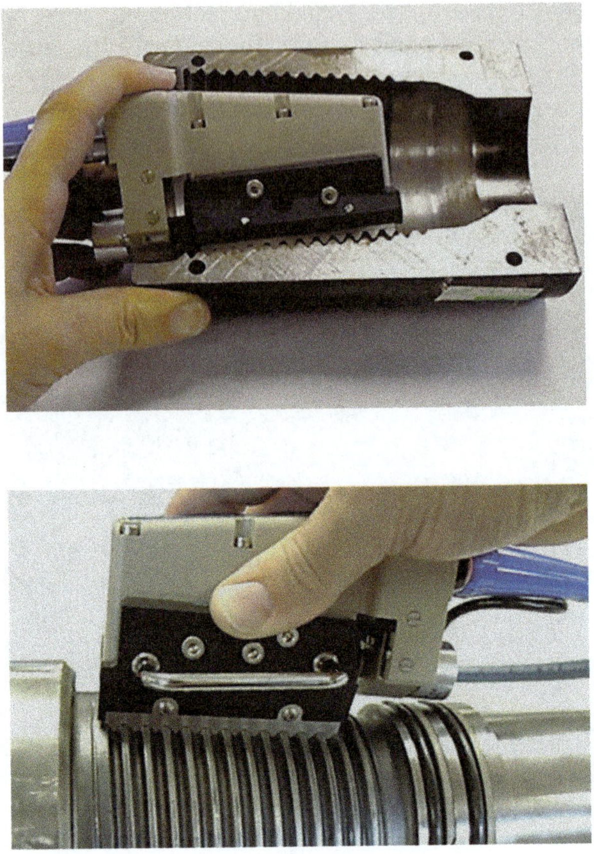

(a)

(b)

Figure 20. Thread array probe: (a) deployed on box; (b) pin connections.

close to, but not in contact with, the thread. The array probes inspect 8 or 12 (depending on the probe tpi) threads in one revolution of the probe.

Where small numbers of connections are required to be tested or where the thread arrays are not suitable, a single-sensor thread probe is used. This probe utilizes plastic shoes tailored for a particular thread type, as shown in Figure 21b. The single-sensor thread probes inspect one thread root with each rotation of the probe; multiple rotations are required to inspect the full thread. They can be used with shoes for both pin and box connections.

Remote Deployed Inspections

The wide coverage of array probes, combined with the relative tolerance to liftoff, enables alternating current field measurement probes to be used with remote deployment solutions, such as scanners and crawlers where access by humans directly is difficult, unsafe, or costly. These applications cover subsea inspection without divers or at a depth where

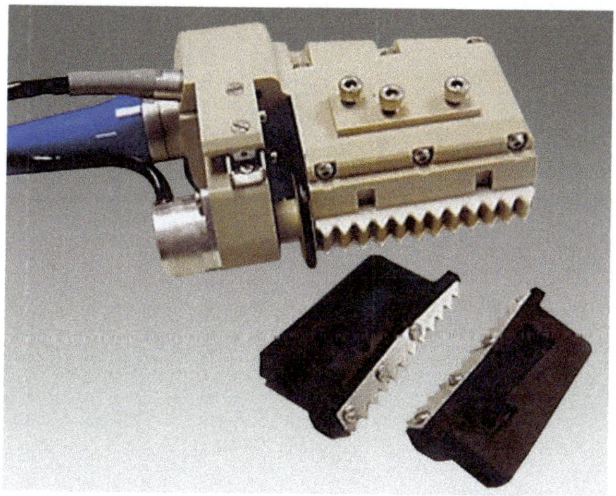

(a)

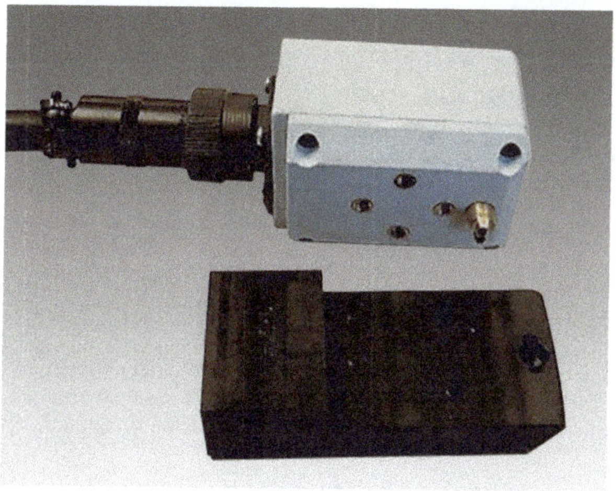

(b)

Figure 21. Thread array probes: (a) with scrapers; (b) single sensor thread probe.

divers cannot operate, as shown in Figure 22. Other applications include topside weld inspections on tall structures.

On carbon steel structures, magnetically attached crawlers are useful deployment tools, with the alternating current field measurement array probe mounted onto a compliant arm, which can keep the probe in contact with the surface (Figure 23). For remote inspection it is essential that cameras, or other means, are used to position the crawler and probe correctly on the structure and to ensure that the probe tracks the inspection zone, such as a weld, within the tolerance of the array.

(a)

(b)

Figure 22. Examples of subsea alternating current field measurement deployment: (a) through magnetic crawler; (b) custom scanner.

Nonferromagnetic Metals

Alternating current field measurement is most often used to inspect ferrous steel structures/components, where the greatest sensitivity is achieved. However, it has also been deployed on nonferromagnetic metals such as austenitic stainless steel (nuclear storage tank pools), titanium (offshore risers), and Inconel™ (threaded connections on down-hole motors).

The signals observed are qualitatively the same as the examples given in the rest of this chapter, albeit with reduced amplitudes, except in the case of a transverse discontinuity. In this case, there is no flux leakage effect, so detection of a transverse discontinuity in a nonferromagnetic metal requires the use of a twin field, or extra scans with the probe rotated by 90°.

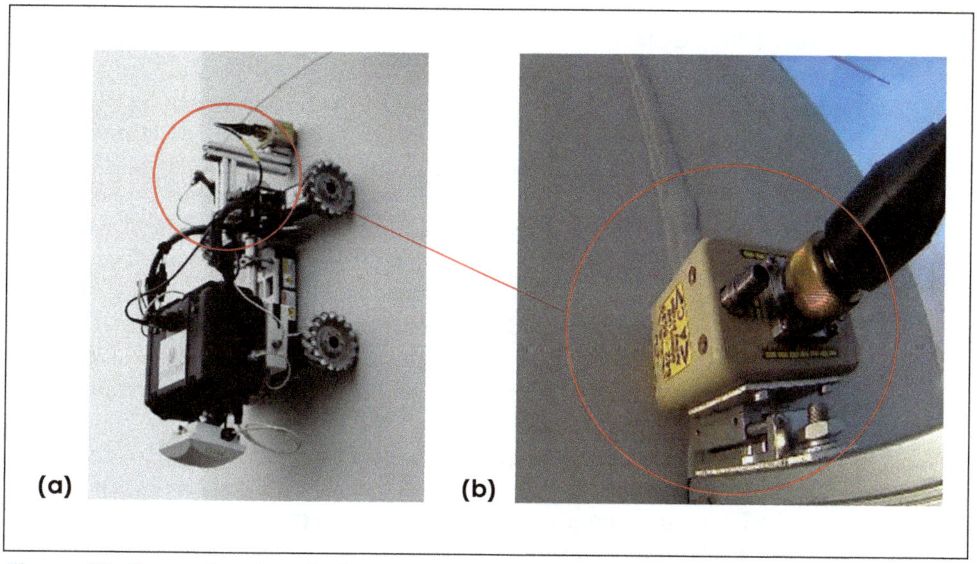

Figure 23. Example of a wireless magnetic crawler deploying an alternating current field measurement array probe: (a) crawler unit; (b) enlargement showing probe.

Acceptance and Standards

Weld Inspection

Alternating current field measurement has been accepted for weld inspection by major organizations primarily in the oil and gas, petrochemical, and marine industries. Approvals take different forms; sometimes they are issued by the asset owner, insurance companies, or certification bodies.

For the oil and gas industry, a number of the certifying authorities have either been involved directly, or reviewed, the large amount of performance data, and have approved the use of alternating current field measurement for the inspection of offshore installations for above-water and subsea applications. These include:

- **Lloyds Register:** *Acceptance of the TSC ACFM Inspection Technique*, 2015.
- **ABS:** *Guide for Nondestructive Inspection of Hull Welds*, 2014.
- **DNV-GL:** *Standard for Certification No. 2.9 Approval Programme No. 402B, October 2008, Firms Engaged in Non Destructive Testing (NDT) on Offshore Projects and Offshore Units/Components.*

In addition to these, documents of recommended practice (or procedures for use) for (or including) alternating current field measurement have been issued by a number of organizations including:

- ***ASTM E2261-03;* latest revision,** *E2261-12: Standard Practice for Examination of Welds Using the Alternating Current Field Measurement Technique.*
- ***ASME Boiler & Pressure Vessel Code,* Section V, Article 15:** *Alternating Current Field Measurement Technique (ACFMT)*, 2006.

Thread Inspection

Alternating current field measurement has been accepted as an alternative to MT for drill-string thread inspection in North Sea standard *NS-2* (1999), latest revision 2012. In addition, a document of recommended practice for alternating current field measurement has been issued:

- ***ASTM E2928-13:*** *Standard Practice for Examination of Drillstring Threads Using the Alternating Current Field Measurement Technique.*

Other Regional Bodies/Organizations

A number of regional bodies and organizations also recognize and/ or approve alternating current field measurement as an inspection technique for welds including:

- **Australia:** alternating current field measurement is included in the National Association of Testing Authorities, Australia (NATA) accreditation scheme and so is nationally recognized as an approved weld inspection technique.
- **Brazil:** Petrobras set up an alternating current field measurement operator certification scheme with training under the Brazilian Association of Non-Destructive Testing and Inspection (ABENDI) scheme.

13

Remote Field Testing

History

The remote field effect was first noted in the 1940s and a system and method were patented by W.R. MacLean in 1951. In the late 1950s, Thomas R. Schmidt independently discovered the technique while developing a tool for testing oil well casings. Schmidt spearheaded the development of the technique and named it remote field eddy current testing. The technique used by the industry is now referred to as remote field testing (RFT). Several test equipment manufacturers recognized the value of this nondestructive electromagnetic technique for the testing of ferrous heat exchanger tubes and began manufacturing remote field test instruments starting in 1985.

Developments since 1990 include the testing of flat plates and steel pipes with external probes that use a technique similar to RFT. Modern instruments use computers to display and store data, and more advanced systems also have automated signal analysis.

Theory of Operation

Remote field testing is a nondestructive testing (NDT) electromagnetic technique that uses *through-wall transmission*. This effect produces a resultant field that is affected by anomalies. It is measured approximately twice the tube diameter away from the excitation source, without any tube magnetization or saturation.

The RFT probe, consisting of an *exciter coil* and one or more *detectors*, is pulled through the tube or pipe. The exciter coil and the detector coil are mounted with an axial distance of approximately two tube diameters (2D) or more between them. The exciter coil is driven with a relatively low-frequency alternating current to produce a changing magnetic field.

This changing primary magnetic field induces strong circumferential eddy currents that extend axially as well as radially in the tube wall. These eddy currents in turn produce their own magnetic field that opposes the magnetic field from the exciter coil. Because the tube wall is ferromagnetic, the magnetic field travels in tight loops near the exciter, within the tube wall. However, at distances between one and two tube diameters (1D and 2D), the magnetic field lines change direction and travel far down the tube before eventually looping back to the other side of the exciter in the space outside the tube.

The exciter coil magnetic field is dominant near the exciter coil and the eddy current magnetic field becomes dominant at some distance away from the exciter coil. The receiving coils are placed where they are unaffected by the magnetic field from the exciter coil but still adequately measure the field that is in the tube wall at distances from the exciter coil of 2D or more.

Electromagnetic induction occurs as the changing magnetic field cuts across the *pickup coils.* By monitoring the consistency of the voltage induced in the pickup coils, changes in the test object can be detected. Although the strength of the magnetic field at a distance from the excitation coil becomes very weak, it remains sensitive to changes in the tube or pipe wall thickness.

Remote Field Testing Zones

Three zones have been identified based on the distribution of the exciter (driver) coil's flux density as it travels down the tube wall:

- Near-field zone: 0 to 1.5 tube diameters from the exciter coil. Also referred to as the *direct-coupled zone*, this is the region where the magnetic field from the exciter coil interacts with the tube wall to produce a concentrated field of eddy currents. This zone does not contribute a great deal of useful data to RFT.
- Transition zone: 1.5 to 2 tube diameters from the exciter coil. The transition zone is the region just outside the near-field or direct-coupled zone. In this zone, the magnetic field in the tube wall changes direction and measurements taken here are not predictable. Its location can change with inspection frequency, wall thickness, relative magnetic permeability, and, to a lesser degree, electrical conductivity.

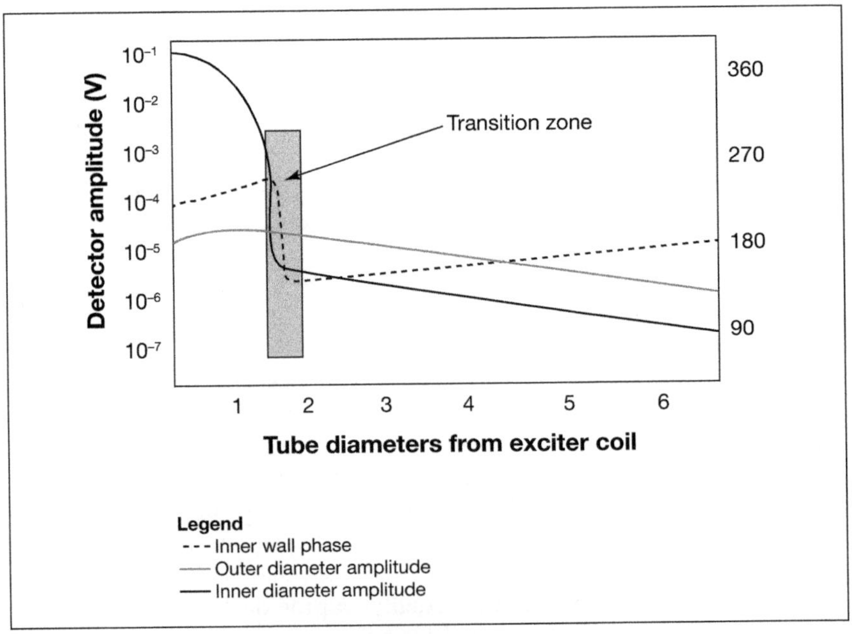

Figure 1. Variations in eddy current density in RFT.

- Remote field zone: 2 to 3 tube diameters from the exciter coil. The outside diameter (OD) field strength is much greater than the inside diameter (ID) field. The induced voltage from a pickup coil is in the microvolt range.

Remote field testing is sensitive to changes in material that occur throughout the entire tube wall thickness if proper frequency selection processes are used. To maintain adequate sensitivity to discontinuities, detectors or pickup coils must be placed in the remote field zone.

Figure 1 shows experimental and model predictions of the detector coil signal magnitude and phase as a function of the distance between the excitation and detector coils expressed in multiples of the pipe diameter. The phase measurements are made with reference to the excitation signal. The magnitude of the detector signal decreases rapidly with distance until a transition region is reached after approximately two pipe diameters in length. Beyond this point, the rate of decrease is much slower. The phase angle, in contrast, changes rapidly in the transition region.

Probe Configuration

A basic remote field probe contains one exciter coil and one receiving (detector) coil. Both coils are wound circumferentially with respect to the tested tube and are separated by a distance greater than twice the tube diameter. If the exciter and receiver were to be placed close together, the detector would measure only the field generated by the exciter in its vicinity. A range of RFT probe types can be seen in Figure 2.

To observe RFT's unique through-wall transmission effect, the technician must move the detector away from the exciter. The detector

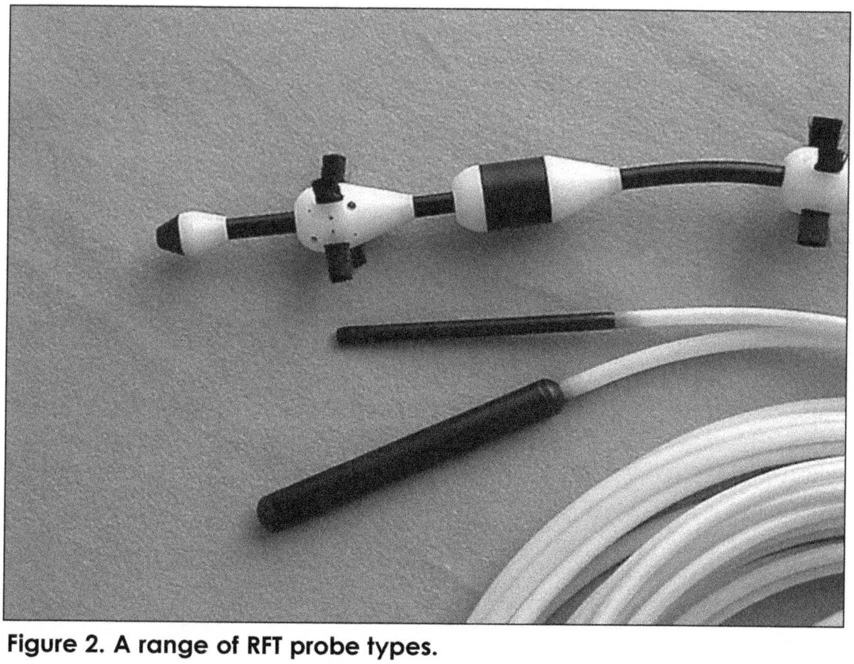

Figure 2. A range of RFT probe types.

measures the electromagnetic field remote from the exciter. The fill factor of the coils can be in the range of 70% or less in some situations. However, keeping fill factors as high as practical is always a better approach. In comparison, most eddy current testing of thin-walled tubing is performed, by code, with fill factors no less than 80%. A lower fill factor reduces sensitivity to small discontinuities but does not affect the quality of RFT data. The ability to function with a lower fill factor makes RFT attractive for pipes with internal coating and tight bends.

Hybrid (driver/pickup) coils are needed to perform RFT. Probe designs are selected based on the number and placement of driver coils, number and style of pickup coils, and overall size based on application requirements.

Probes for the inspection of tubes or pipes are typically of the bobbin (inside) type. These probes can use either single or dual excitation or detection coils. The excitation coils are typically driven by low-frequency alternating currents (50 Hz to 1 kHz).

Probe pickup coils can be used in absolute or differential modes for detection of a wide range of wall loss conditions that may be either ID or OD in origin.

- The single-coil pickup is known as an *absolute coil* because its voltage output is directly proportional to the local value of the magnetic flux density. Absolute coils are excellent for sizing large-volume discontinuities, such as general wall loss and erosion.
- A *differential detector* is formed by wiring two adjacent absolute coils with opposing outputs. The differential detector subtracts out slow variations in magnetic flux density and produces a larger response to sharp changes in the field, such as those occurring at pits. Differential coils can also be used to size large-volume metal loss.

Both absolute and differential coils can be split into a set of smaller coils called *array detector probes*. Each coil in the array is sensitive to the adjacent segment of the tube circumference. Array probes tend to be more sensitive to small discontinuities and they also produce data that indicate the circumferential extent of metal loss from the response of each coil.

Effects of Probe Speed

The ability of RFT to detect and quantify discontinuities relies on the quality of the signal received. The probe must be in the vicinity of the discontinuity for at least one excitation cycle in order to detect it. The probe pull speed should be slow enough so that the digital sample rate allows the field profile near the probe to be accurately recorded. It is equally important that the speed of testing be as constant as possible. Sudden changes in speed can result in anomalous signals. For most work in heavier wall tubing greater than 0.05 in. (1.3 mm) with frequencies below 1 kHz, a practical pull speed will be in the range of 6 in. (15 cm) per second.

Frequency Selection

Remote field testing operates at relatively low frequencies typically ranging from 40 to 500 Hz. In general, a lower frequency (up to 250 Hz) is used for thick-walled and high-permeability pipe. For higher sensitivity and high test speeds, a higher frequency is used. The frequency is chosen as high as possible while minimizing noise and remaining in the remote field zone. It is important to check for the presence of any electromagnetic noise sources such as welders, electric motors, and pumps that tend to generate noise in the frequency range of RFT.

Features of Remote Field Testing

Sensitivity

The accuracy for RFT in the straight part of tubes is about 5% of wall thickness for general wall loss. The accuracy is less (20% of wall) for highly localized discontinuities and in bends or near external conducting objects because of the changes in magnetic properties of the tube in the bend area and due to shielding effects of external objects.

Remote field testing can detect both inside and outside surface discontinuities with equal sensitivity, but in most cases cannot differentiate between them without using near field coils. Remote field testing is insensitive to scale and nonferromagnetic debris. Internal and external ferromagnetic deposits do affect the signal to a small degree.

As noted above, a large fill factor is not required for RFT, and centralization is not as critical as with other NDT techniques for similar applications.

Signal Analysis and Data Presentation

Remote field test data are recorded in computer memory or hard drive. Phase-amplitude diagrams (voltage planes) are displayed during the test on instrument monitors in real time. The raw data from the detector are stored either in phase amplitude format or as in-phase and quadrature components. The data can be recalled for display, analysis, and final report preparation (Figure 3).

The voltage plane and x,y displays provide maps of the detector coil output in polar coordinates. On polar displays, signals are drawn as vector points with the angle representing the phase and the radius representing the amplitude. Remote field testing signals on the voltage plane (Figure 4) or x,y display (Figure 5) are scaled and rotated to a convenient position for final viewing.

By observing the signal shape, phase angle, and relative size on the voltage plane, many discontinuities can be characterized and sized for depth and circumferential extent. The axial length of a discontinuity can also be measured by recording the data on a strip chart as the probe is pulled through the tube. All remote field test instruments are able to display the data as strip charts and voltage planes. In addition, some instruments have automatic depth sizing and reporting software.

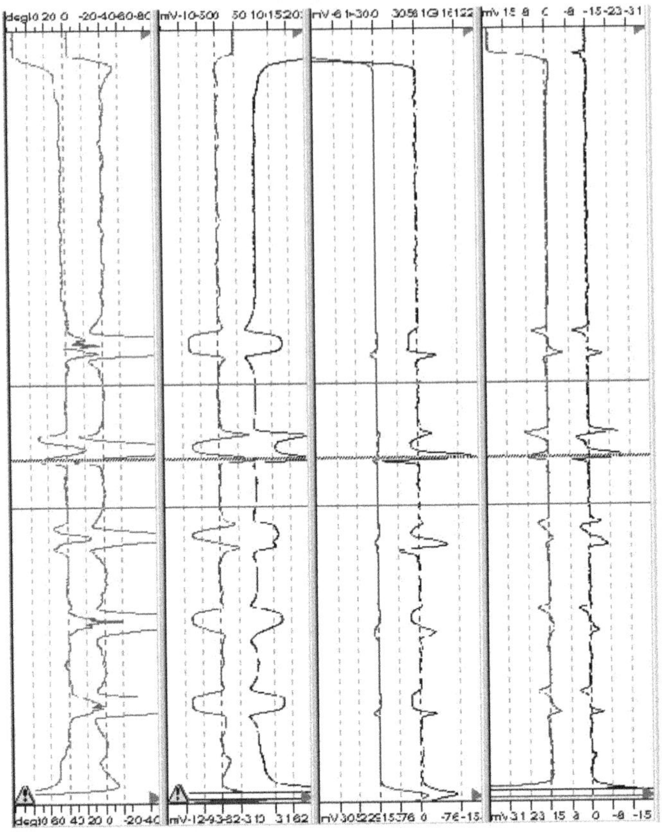

Figure 3. Strip chart recordings: (a) phase and log amplitude signals for absolute probe; (b) x,y voltage signals; (c) differential signals; and (d) mixed-frequency signals.

Reference Standards

All NDT methods use reference standards to compare discontinuity signals with those from known machined or artificial discontinuities. Remote field testing requires reference standards for each variation in tube diameter, wall thickness, conductivity, and permeability. Reference discontinuities must be machined to closely simulate the discontinuities expected in the tube or pipe being examined. *ASTM E2096-16: Standard Practice for In Situ Examination of Ferromagnetic Heat-Exchanger Tubes Using Remote Field Testing,* lists two recommended reference standards that can be used for RFT.

Remote Field Testing Applications

Remote field testing can be used for all conventional carbon steel material specifications, diameters, and wall thicknesses. The test speed

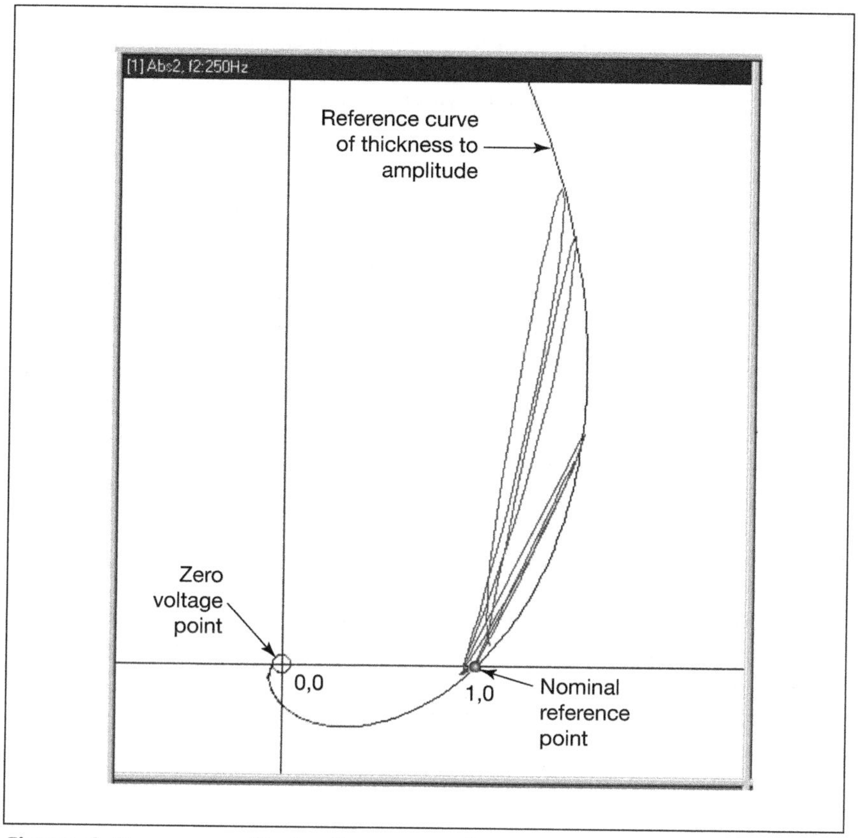

Figure 4. Phase-amplitude diagram; voltage plane recording.

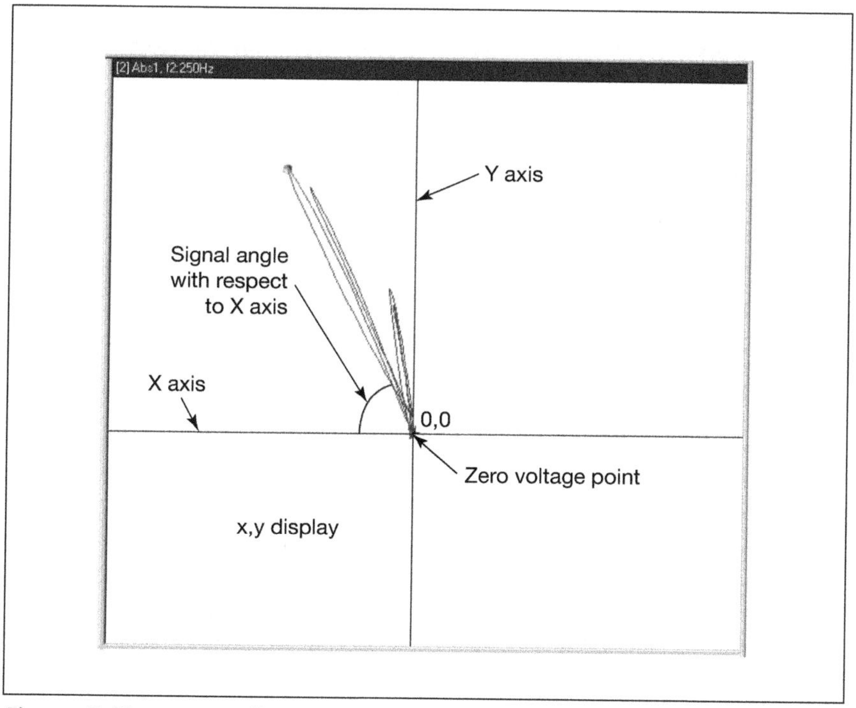

Figure 5. Phase-amplitude diagram; x,y plane recording.

can be up to 60 ft (18 m) per minute depending on wall thickness and test frequency. Remote field testing is a noncontact technique. The probes have minimal friction with the tube or pipe wall and require no couplant.

The use of the multipurpose standard defined in *ASTM E2096* is to verify that at the time of the test the equipment and probe used are meeting a specific sensitivity level. The system must demonstrate that it can detect the known discontinuity conditions modeled on the reference tube. It must be able to detect both small-volume (pit-like) events as well as larger-volume erosion corrosion. This initial standardization should be recorded as part of the permanent test record.

In an eddy current test of nonferritic tubing, once the test configuration has been decided and the tester values set, they will never be adjusted during the examination of the heat exchanger. In RFT, that is not the case. It is almost guaranteed that the operating frequency will need to be changed during the test to compensate for shifts in material conditions. If this is not done correctly, the test is of no value.

During the standardization process, the technician notes the screen orientation of certain signal sources on the standard (Table 1). This review is normally done from the differential channel. If the proper test frequency has been selected:

- The ASTM through-wall hole will lie at approximately 90° on the screen.
- The 20% OD groove will lie at approximately 155° on the screen.
- The simulated support plate element may have a range of "typical" responses.

Table 1. Signal rotation settings for tube support thicknesses

Tube support thickness	Signal rotation setting
0.125 in. (3.2 mm)	0–4°
0.250 in. (6.4 mm)	5–10°
0.375 in. (9.5 mm)	≈ 15°
0.500 in. (12.7 mm)	≈ 20°
0.625 in. (15.9 mm)	≈ 25°
0.750 in. (19 mm)	≈ 30°
0.875 in. (22.2 mm)	≈ 35°

The thinner supports will lie lower on the horizon and have much less amplitude (voltage). As the baffles get thicker, the responses seen will rotate higher above the horizon and will increase in amplitude (Figure 6).

Standardization curves can be established on each channel using phase-angle-to-discontinuity depth, voltage-to-discontinuity depth, or some combination of the two. To initially verify the full range of discontinuities, it is sometimes useful to look at some additional large-volume discontinuity information. There are standards that use what are called 180 and 360° discontinuities to do this (Figure 7).

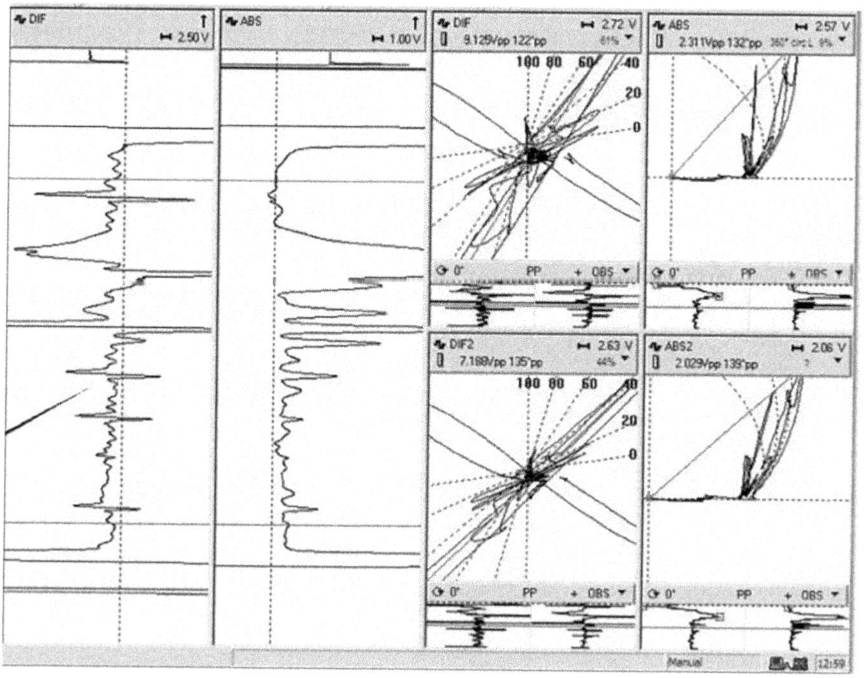

Figure 6. Responses seen after the standardization process has been set up from a two-frequency test viewing both the differential and absolute channels. The full range of discontinuity types and depths as well as the support element are shown.

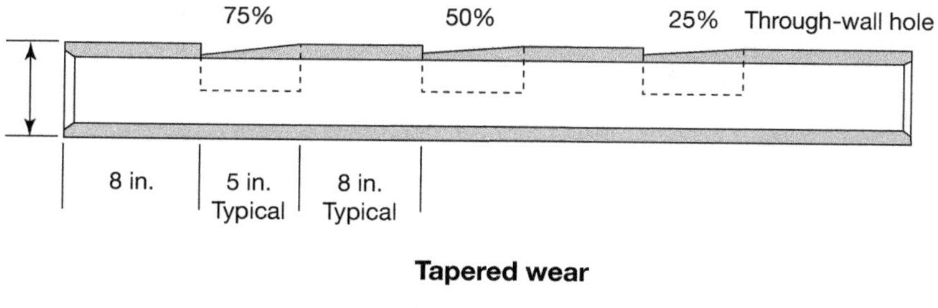

Figure 7. 180° tapered wear standard.

By using these three types of standards, it is possible to build and verify the precision of the system to more accurately estimate tube wall loss from a wider range of damage mechanisms.

At this point, the system has been proven to be working correctly. Standardization curves (Figure 8) have been created to help the data analyst quantify any damage seen in the heat exchanger. The probe is inserted into a tube in the center of the heat exchanger, and the data are reviewed as the probe moves at the selected inspection pull speed.

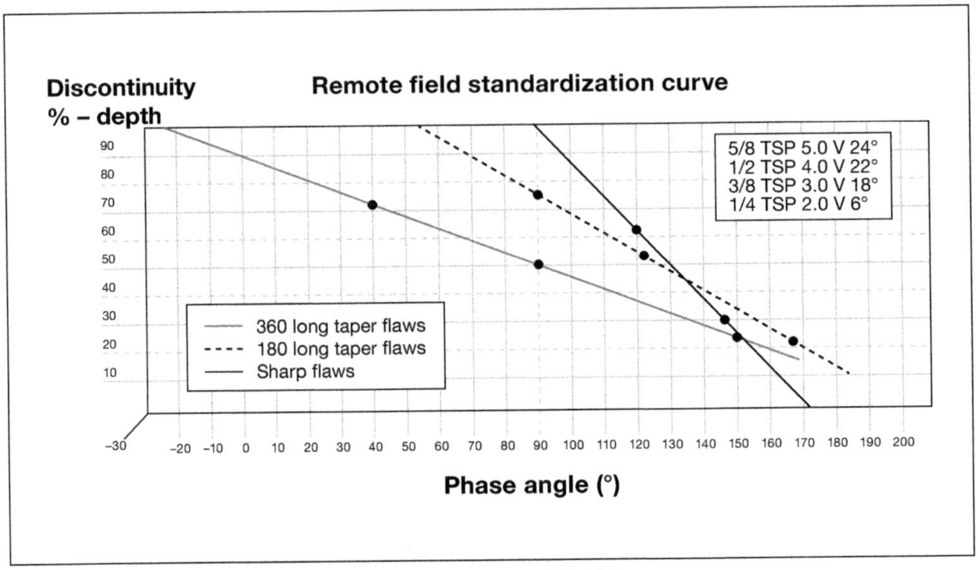

Figure 8. Theoretical pit curve, 180° curve, and 360° curve.

The first item that is noted is that the supports are no longer falling at the expected phase angle as they are seen on the test output screen. It is now necessary to adjust the test frequency to get back to the phase angle orientation and relationships as were established on the reference standards. The elements from the tube bundle that are used to evaluate this are:

- The tube support plate response.
- The tube end exit signal (TEES) for massive through-wall holes.

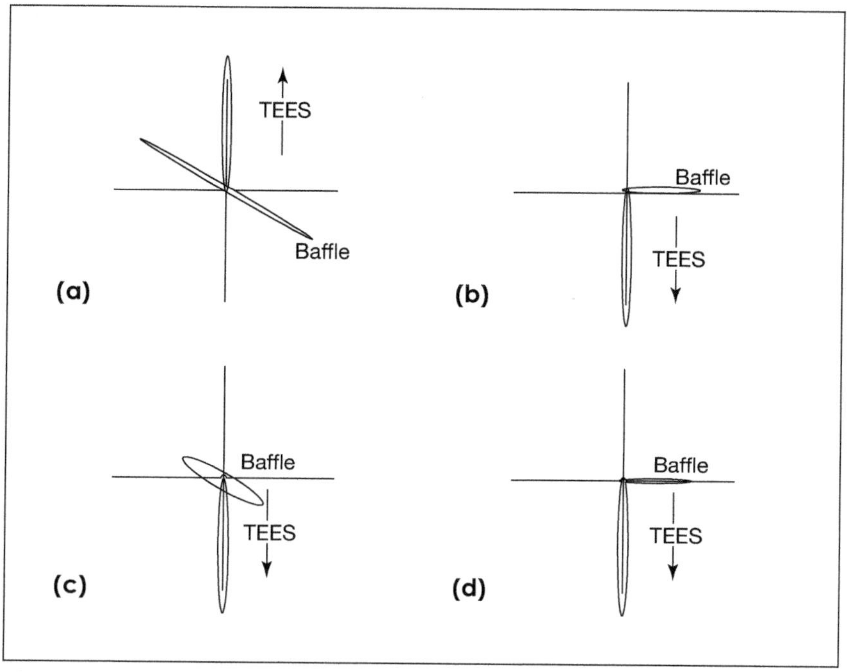

Figure 9. Theoretical heat exchanger standardization set points: (a) differential/dual exciter; (b) absolute/dual exciter; (c) differential/single exciter; (d) absolute/single exciter.

The technician adjusts the test frequency to the range where phase angle ratios are brought back to their former positions (Figure 9). Once this frequency adjustment takes place, the original standardization curves can be used as initially produced.

14

Magnetic Flux Leakage Testing

Factors Affecting Flux Leakage Fields

Discontinuity Geometry, Location, and Orientation

The presence of a discontinuity causes a reduction in the cross-sectional area of the test object, thereby resulting in a local increase in the magnetic flux density. A reduction in the permeability, together with an increase in the magnetic flux density, causes the flux to leak into the surrounding medium.

Magnetic leakage fields can be subdivided further into active or residual leakage fields. To understand the origin of the leakage fields and choice of initial magnetization for the active leakage field technique, consider an unmagnetized steel billet with a surface discontinuity, as shown in Figure 1a.

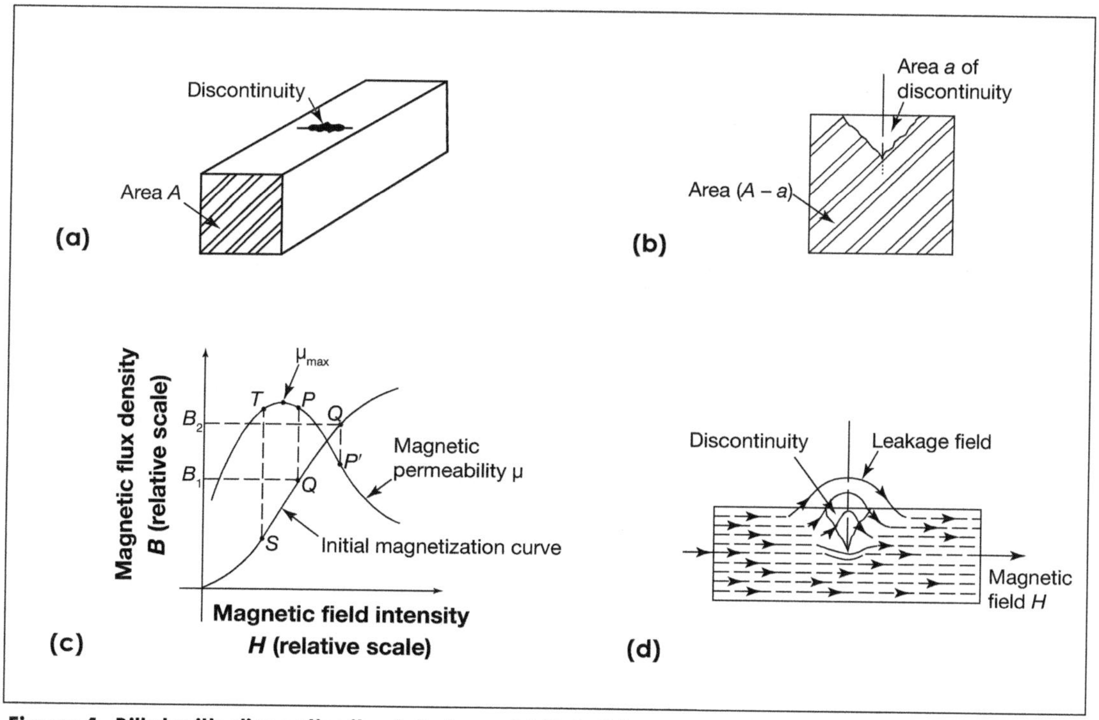

Figure 1. Billet with discontinuity: (a) view of billet; (b) cross section through discontinuity; (c) magnetic characteristics of billet material; and (d) billet in magnetic field showing discontinuity leakage field.

Let A represent the cross-sectional area of the billet, and let a (Figure 1b) represent the cross-sectional area of the discontinuity. The cross-sectional area of the sound portion of the billet in the vicinity of the discontinuity is reduced to $(A - a)$ units.

H, illustrated in Figure 1c, is the magnitude component and therefore the scalar quantity. With the billet placed in a uniform magnetic field H, the induced flux density in the sound portion of the billet is represented by B_1 (W/m^2). This magnetic flux density corresponds to point P to the right of μ_{max} on the permeability curve of the material.

Point Q is the corresponding point on the initial magnetization curve in Figure 1c. If it is assumed that the same magnetic flux from the sound portion of the billet (B_1) is to pass through the reduced billet area in the vicinity of the discontinuity, then the flux density present in this section is greater than $B_1 \times A\ (A - a)$, namely B_2. This local increase of magnetic flux density results in a change of the operating point on the magnetization curve from Q to Q' and the corresponding decrease of local permeability from P to P'. However, this results in conflicting demands in the vicinity of the discontinuity.

The magnetic flux density must increase with the reduction of the cross-sectional area. This change drives the permeability in the restricted region of the billet to a value of less than that present in the sound regions. Consequently, some of the flux leaks into the surrounding medium near the discontinuity, which is why this is called a leakage field, as illustrated in Figure 1d. The detection of this leakage field is the basis for magnetic flux leakage testing (MFL).

Subsurface Discontinuities

If a discontinuity is below the surface, the difficulty of detecting the magnetic leakage field is much greater. The reason for this difficulty is that the surrounding material tends to smooth out the field distortion caused by the subsurface discontinuity, thus resulting in a small field disturbance on the surface of the billet. Because some detectors (for example, those based on coils) can only detect relatively sharp changes of the magnetic flux leakage field, it is difficult to sense the location of subsurface or lower surface discontinuities (Figure 2). This may not be

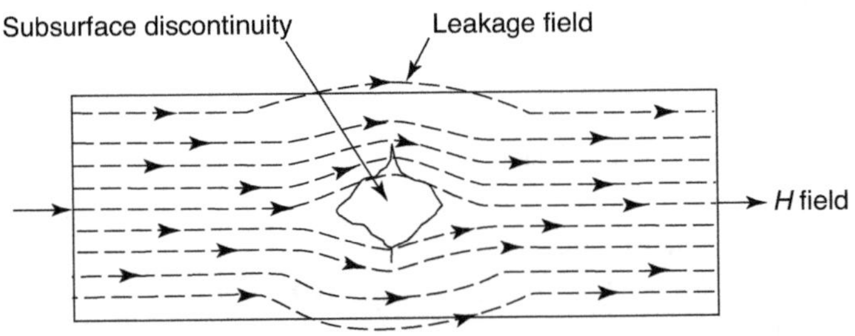

Figure 2. Billet with subsurface discontinuity showing resultant leakage field.

the case for sensors (for example, hall effect or magnetodiode sensors) that rely on the absolute value of at least one component of the leakage field rather than its gradient, that is, how fast the magnetic flux leakage changes as the sensor passes through it.

Degree of Initial Magnetization

The initial operating point on the permeability characteristic of the material is very important. If the point is to the left of μ_{max} (point *T* in Figure 1c), an area reduction caused by a discontinuity will drive the local permeability higher than the permeability of a material free of discontinuities. Thus, there is a possibility that the discontinuity may go undetected. Moreover, if the initial magnetization of the material should locate the operating point near saturation, then the difference between the magnetic flux density in the material and the leakage magnetic field in the surrounding medium decreases with the increasing discontinuity cross-sectional area.

Therefore, the difficulty of detecting the discontinuities is magnified because it becomes increasingly difficult to discriminate between the severities of the various heterogeneities. Also, because the degree of magnetization is so great, the surface roughness is easily mistaken for actual discontinuities and results in unwarranted rejection of the test objects. Thus, there exists an upper and lower limit of magnetization to which a test object should be subjected if MFL is to be successful. Magnetization of the test object should lie on the downslope of the μ versus *H* curve.

In active magnetization (inspection while the exciting field is on), the magnetization should be where the *B-H* curve is flattening. Too high a magnetization may result in excessive magnetic noise, and too low may result in minimal magnetic flux leakage from imperfections, so that they also become hidden in the noise.

For much MFL where upper, mid-wall, and lower surfaces of parts may contain imperfections (for example, when inspecting steel pipe or tank floors), the material's magnetization point should be at least where the *B-H* curve is flattening out. This level will generate magnetic flux leakage from subsurface imperfections, especially those with relatively large volumes, such as pits.

Inspection in Residual Induction

Tight surface discontinuity detection is performed using residual induction after magnetization by one or more "shots." Here also, it is essential that the magnetizing shot(s) take the material to a high point on the *B-H* curve so that the resulting residual induction is close to the *remanence value* for the material. If it is too low, the resulting magnetization may cause no perceptible magnetic flux leakage from imperfections, even where the sensor liftoff is effectively zero.

Steps in Magnetic Flux Leakage Testing

There are four steps in MFL:
1. Magnetize the test object so that discontinuities perturb the flux.
2. Scan the surface of the test object with a detector sensitive to magnetic flux.
3. Process the raw data from these detectors in a manner that best accentuates discontinuity signals.
4. Present the test results clearly for interpretation.

Selection of Magnetization Method

Successful testing requires the test object to be magnetized properly. This can be accomplished using one of three approaches:
- Permanent magnet.
- Electromagnet.
- Magnetizing coil.

Permanent Magnets

Excitation systems that use permanent magnets offer the least flexibility. Such systems use high-energy permanent magnet materials such as neodymium iron-boron, samarium-cobalt, and aluminum-nickel.

The major disadvantage with such systems is that the excitation cannot be switched off. Because magnetization is always present, it is difficult to insert and remove the test object from the test instrument. Although the magnetization level can be adjusted using appropriate magnetic shunts, it is awkward to do so. Permanent magnets, typically neodymium iron-boron, have found increasing use as an inline tool (called *smart pigs*) for pipeline inspection, coiled tubing inspection systems, tank floor inspection systems, and wire rope inspection systems where the high currents occurring in magnetizing coils are not allowed for safety reasons.

Electromagnets

Electromagnets are used extensively to magnetize test objects. Figure 3 shows a system where current passes through an excitation coil. This makes the test object part of a magnetic circuit energized by the current.

When the coil wound on the yoke carries current, the resulting magnetic force drives magnetic flux through the yoke and the test object, between the poles of the yoke. The total magnetic flux $\hat{E}$ (in webers) is given by:

(Eq. 1) $$\hat{E} = \frac{NI}{S}$$

where
I is current (amperes) in the coil
N is the number of turns in the coil
S is the reluctance (amperes per weber) of the magnetic circuit

To obtain maximum sensitivity, it is necessary to ensure that the magnetic flux is perpendicular to the discontinuity. Because the orientation of the discontinuity is unknown, it is necessary to test twice with the yoke, in two directions perpendicular to each other. A grid is usually drawn on the test object to facilitate the tests.

In some modern yoke systems, the coil (shown in Figure 3) is replaced by neodymium iron-boron permanent magnets. This is used for testing in wire rope, coiled tubing, and tank floor systems.

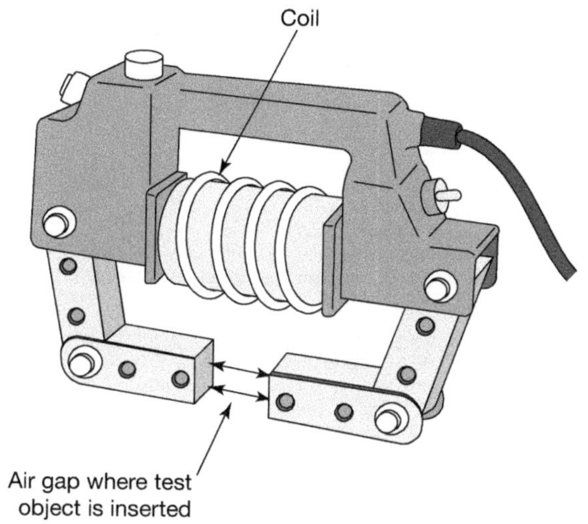

Figure 3. Electromagnetic yoke for magnetizing a test object.

Magnetizing Coils

A commonly used encircling coil is shown in Figure 4. The field direction follows the right-hand rule. As presented earlier, the right-hand rule states that if someone grips a rod, holds it, and imagines an electric current flowing with the thumb, the induced circular field in the rod would flow in the direction that the fingers point. With no test object present, the field lines form closed loops that encircle the current-carrying conductors. The value of the field at any point has been established for a great many coil configurations. The value depends on the current in the coils, the number of turns (N), and a geometrical factor. Calculation of the field is generally unnecessary for nondestructive testing, since a hall effect sensor will measure this field for the inspector.

The introduction of the test object into the field of the coil changes the field. The metal becomes part of the magnetic circuit, with the result that close to the surface of the test object (at R in Figure 4), the magnetic field intensity (H) is lower than it would be if the test object were removed. A hall effect sensor will show the field intensity at the test object is not the same.

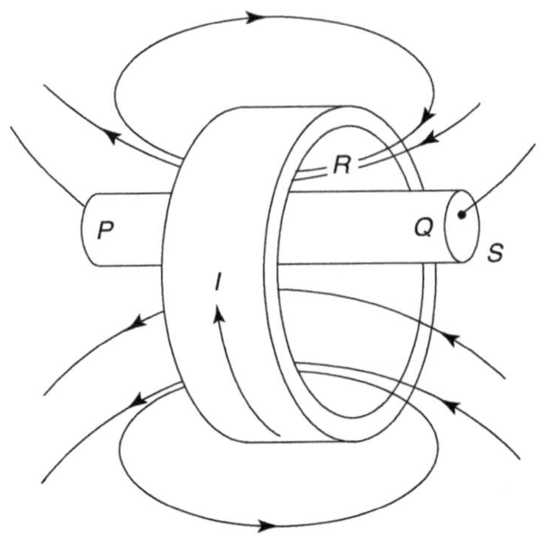

Legend
I = electric current
P, *Q* = points of discontinuities in test object
R = point at which magnetic field intensity *H* is measured
S = point at which magnetic flux density *B* is measured

Figure 4. Encircling coil using direct current to produce magnetizing force.

Testing in Residual Field

Test objects are first passed through the coil field and then tested in the resulting residual field. Elongating the coil and placing the test object next to the inside surface of the coil exposes the test object to the largest field that the coil can produce. (This technique is often used in magnetic particle testing.) The technician should experiment to optimize the coil field requirements for the test object because this field depends on the test object geometry.

Magnetizing by Applied Direct Current

If an electric current is used to magnetize the test object, it may be more advantageous to orient the direction of current in a manner where the presence of a discontinuity impedes the current flow as much as possible. Bars, billets, and tubes are often magnetized by applying direct current to their ends, as shown in Figure 5.

Another magnetizing technique using direct current includes passing the current directly through a tubular test object to magnetize the test object circularly, as illustrated in Figure 6.

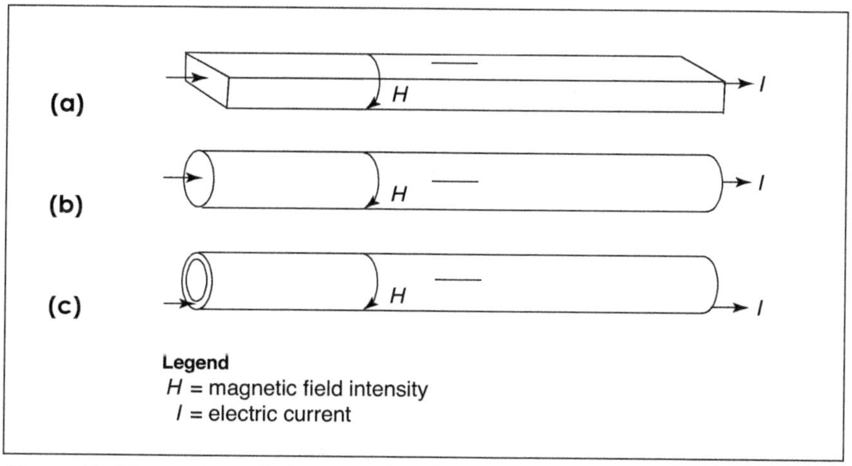

Figure 5. Circumferential magnetization by application of direct current: (a) rectilinear bar; (b) round bar; and (c) tube.

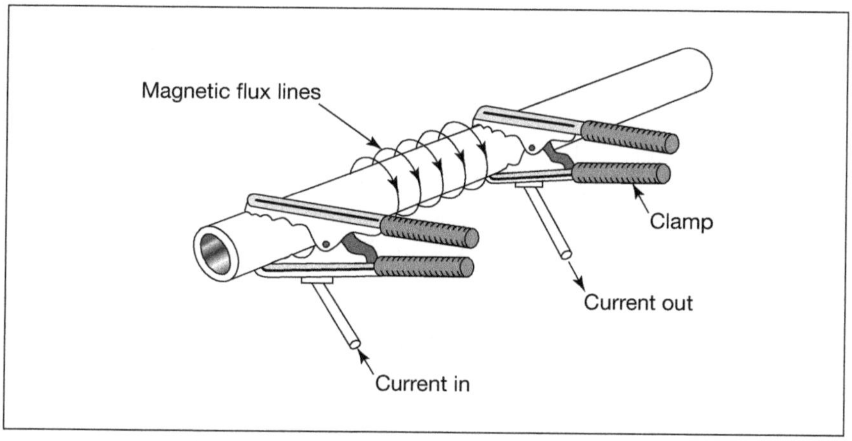

Figure 6. Current-carrying clamp electrodes used for testing ferromagnetic tubular objects with small diameters.

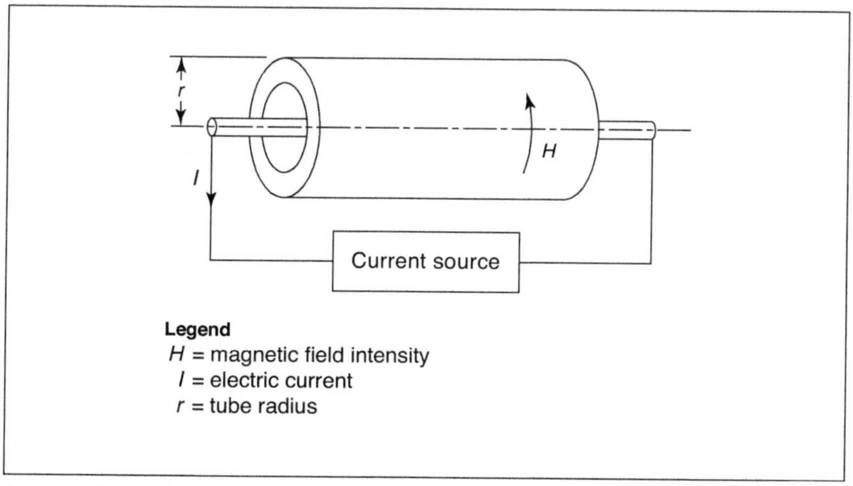

Figure 7. Central conductor using direct current to produce magnetizing force.

Figure 7 shows a central conductor energized by a current source (I), again to establish a circular magnetic field intensity H (amperes per meter) in a tubular test object:

(Eq. 2)
$$H = \frac{I}{2\pi r}$$

where
I is direct or peak current in amperes on long shots (for example, 0.5 s)
r is distance from center of rod to the outer edge of the tube (m)

As the length of the shot shortens to a pulse, such as when capacitor discharge systems are used as the current source, Equation 2 ceases to be applicable. This is because the rapid rise of current creates eddy currents in the tube, which hinder penetration of the magnetizing field throughout the shot.

Capacitor Discharge Devices
For the circular magnetization of tubes or the longitudinal magnetization of the ends of elongated test objects, a capacitor discharge device is sometimes used. The capacitor discharge unit represents a practical advance over battery packs and consists of a capacitor bank charged to a voltage (V) and then discharged through a rod, a cable, and a silicon-controlled rectifier of total resistance (R).

Magnitudes of Magnetic Flux Leakage Fields
The magnitude of the magnetic flux leakage field under active direct current excitation naturally depends on the applied field. However, in the case of residual magnetization, the magnetic flux leakage fields may be only a few gauss (a few hundred micro-tesla). Furthermore, with residual field excitation, a magnetic leakage field polarity reversal may occur depending on the value of the initial active field excitation and the dimensions of the discontinuity. At the point of reversal, there is no magnetic flux leakage from the discontinuity. To overcome this effect, the magnetizing shot(s) should leave a large amount of flux in the part, and the part should be taken close to magnetic saturation during the magnetization process.

15

Magnetic Flux Leakage Applications

Introduction

Magnetic flux leakage testing (MFL) is a commonly used nondestructive testing method. Signals from probes or sensors are processed electronically and presented in a manner that indicates the presence of discontinuities. Although some MFL techniques may not be as sophisticated as others, it is probable that more ferromagnetic material is tested with MFL than with any other method.

Heat Exchanger Tubing Applications

Magnetic flux leakage testing is widely used to test installed ferromagnetic heat exchanger and boiler tubes. The flux leakage probe consists of a magnet and two inductive coil sensors. The magnets set up a flux field in the tube wall as they pass through the tube. The field fluctuates when it encounters a discontinuity. This fluctuation is picked up by the coils and appears on the display apparatus and chart recorder.

Magnetic flux leakage testing is based on the influence of discontinuities on a magnetic field. The technique is limited to ferromagnetic material. The MFL probe consists of a magnet with two types of magnetic pickups: coil and hall element. The coil picks up the flux rate of change, whereas the hall effect sensor picks up absolute flux. The coil detects small discontinuities that cause perturbations in the flux. Magnetic flux leakage coils are more sensitive to sharp discontinuities than to gradual wall loss. In fact, the coils can totally miss long areas of wall loss if the changes in wall thickness are gradual. Thus, a hall effect sensor is used to detect gradual wall loss but not pits. Figure 1 shows a probe configuration for flux leakage testing.

The output of the hall effect sensor depends on the orientation of the sensor in the probe relative to the discontinuity and whether the location of the discontinuity is on the inside or outside surface. Inside surface discontinuities produce larger signals than outside surface discontinuities because the field strength on the inside surface is higher than on the outside surface.

It should be noted that the output of the MFL coil is related to the change of flux caused by the discontinuity but not directly to discontinuity depth. Tubes up to 3.5 in. (8.9 cm) in diameter and 0.012 in. (0.03 cm) wall thickness can be tested.

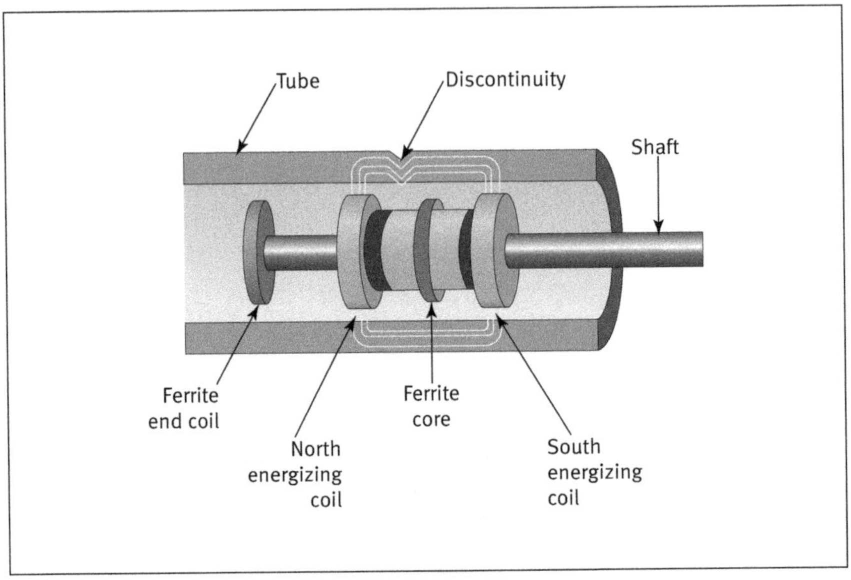

Figure 1. Bobbin probe for MFL.

Figure 2. Electromagnetic testing of mooring rope.

Wire Rope Inspection

Because the reliability of wire ropes is crucial for the safety of many mining, oil industry, crane, and ski lift operations, concern with their integrity is a constant preoccupation of users and safety authorities. Effective procedures, combined with a good understanding of degradation mechanisms and discard criteria, can notably increase wire rope safety.

Testing is performed to determine cross-sectional loss caused by corrosion and wear, and to detect internal and external broken wires. The type of flux loop used (electromagnet or permanent magnet) can depend on the accessibility of the rope. Permanent magnets might be used where taking power to an electromagnet might cause safety problems.

The cross-sectional area of the rope can be measured by sensing changes in the magnetic flux loop that occur when the rope gets thinner. The air gap becomes larger, so the value of the field intensity falls. This change can be easily sensed by placing hall effect sensors anywhere within the magnetic circuit.

Modern dual-function electromagnetic rope testers, shown in Figure 2, allow simultaneous tests for loss of metallic cross-sectional area and localized discontinuities. For this tester, strong permanent magnets induce a magnetic flux at the saturation level in the rope in the axial (longitudinal) direction. Various types of sensors close to the rope (such as coils, hall sensors, or flux gate sensors) sense and measure the magnetic flux. Any discontinuity, such as broken wire or corrosion pitting, distorts the magnetic flux in the rope and causes it to leak from the rope. For localized discontinuity tests, the radial component of the leakage flux is measured by radial sensors.

Round Bars and Tubes

In some test systems, round bars and tubes are magnetized by an alternating current magnet and rotated under the magnet's poles. Because the flux leakage from surface discontinuities is very weak and confined to a small area, the probes must be very sensitive and small. The system uses a differential pair of magnetodiodes to sense leakage flux from the discontinuity.

The differential output of these twin probes is amplified to separate the flux leakage from the background flux. In this system, pipes are fed spirally under the scanning stations to increase the test rate. In one similar system, round billets are rotated by a set of rollers while the billet surface is scanned by a transducer array moving straight along the billet axis. Seamless pipes and tubes are made from the round billets.

Petroleum and Gas Pipelines

Pipelines connect field production (gas and oil extraction) with refineries and petrochemical plants where gas and crude petroleum are processed into usable products. Because pipelines cross state lines in the United States, they are governed by the Department of Transportation. Construction, maintenance, and testing of these pipelines are critical to the safety of the environment and the general public. Buried pipelines not only have the potential for catastrophic failure but could contaminate water sources if leakage occurs.

Magnetic flux leakage testing can be used to detect cracks and corrosion pits in the walls of underground pipelines. In pipe manufacturing facilities, the magnetic flux leakage sensitivity is set using electrical discharge machining (EDM) notches, the depth of which is a specified fraction of the nominal wall thickness. Typically, cracks and hairline cracks, 0.006 in. (0.015 cm) deep on the outside diameter (OD) and 0.012 in. (0.03 cm) deep on the inside diameter (ID), are satisfactorily detected by MFL techniques for wall thicknesses up to 0.315 in. (0.8 cm).

The discontinuities that cause pipeline failures are typically much larger than hairline cracks and thus can be detected more easily in principle. However, the technological difficulties to be overcome are challenging. First, the entire testing device must be transported inside the pipe for long distances. Magnetizing conditions close to saturation must then be achieved to reveal outside diameter discontinuities with inside diameter probes, requiring very power-efficient magnetizing systems. Furthermore, the entire pipeline must be tested with only one pass of the probes and associated magnetizing device. In addition, recording systems must be capable of giving definitive positional information.

For the pipeline inspection tool, a recorder package is added and the signals from discontinuities are recorded. When the recordings are retrieved and played back, the areas of damage are located. Pipe welds provide convenient magnetic markers.

Smart pigs are test vehicles that are pushed through the pipeline by product flow. A diagram is shown in Figure 3a. The technique got its name from the sound of the original scraper pig moving through the pipe. At the end of the line or run, the pig is retrieved and the onboard data are then processed and analyzed, as shown in Figure 3b. The pigs are similar to the MFL probes used in tube testing, but pigs are constructed to be propelled down pipelines and collect the required test data.

Billets

A relatively common problem with square billets is elongated surface-breaking cracks. By magnetizing the billet circumferentially, MFL can be performed to detect such discontinuities.

Magnetic flux leakage systems for testing tubes exhibit the same general ability to classify discontinuity depth. It is generally accepted that even with the lack of correlation between some of the instrument readings and the actual discontinuity depths, the automatic readout of these two systems still represents an improvement over visual or magnetic particle testing.

Aboveground Storage Tank Floors

Tank floors of aboveground storage tanks, shown in Figure 4, are subject to corrosion where they touch the ground. Magnetic flux leakage testing has been widely used in the oil field industry for over 25 years for the examination of pipe, tubing, and casing, both new and used. It is only in the past 10 years that this testing method has been applied to aboveground storage tank floors in an attempt to provide a reliable indication of the overall floor condition within an economical timeframe.

For the purpose of this application, a magnetic bridge is used to introduce as near a saturation of flux as is possible in the test material

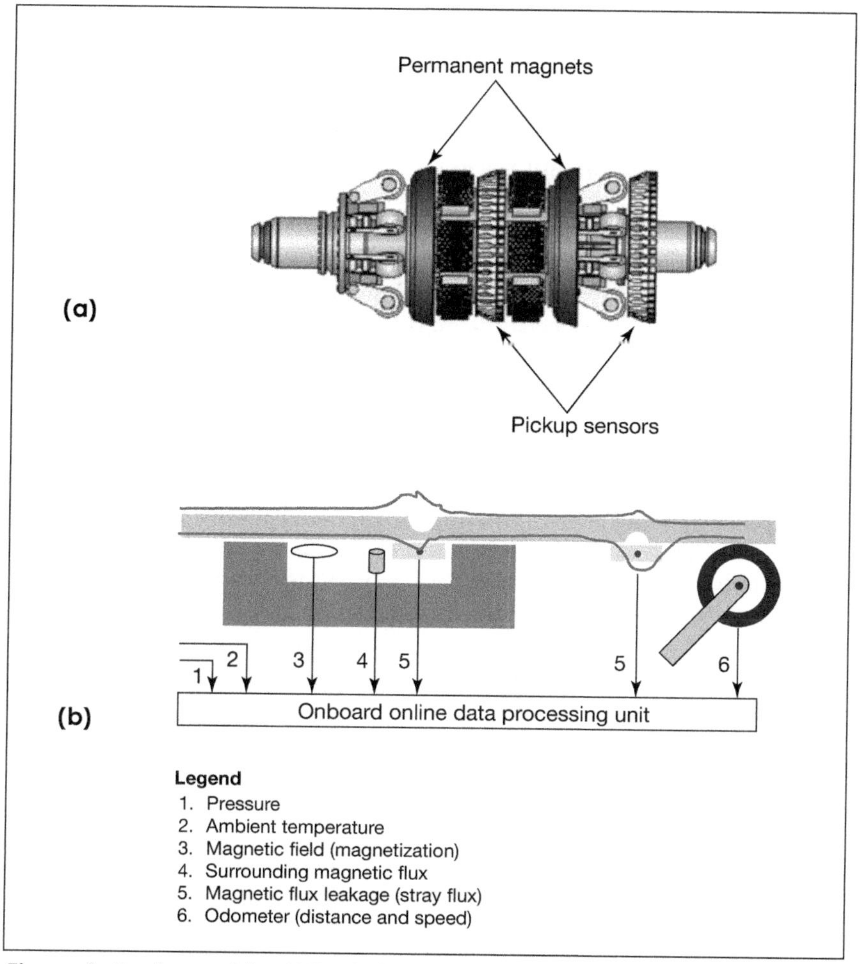

Figure 3. Equipment for MFL of pipes and tubes: (a) smart pig; (b) data acquisition from pig sensors.

Figure 4. Aboveground storage tank for petroleum products.

between the magnetic poles. Any significant reduction in the thickness of the plate will result in some of the magnetic flux being forced into the air around the area of reduction. Flux leakage sensors that can detect these flux leakages are placed between the poles, as diagrammed in Figure 5a. A mobile floor-scanning unit is shown in Figure 5b. It is important that MFL equipment produced for this particular application be designed to handle the environmental and practical problems always present.

Powerful rare earth magnets are ideally suited for this application. They are more than capable of introducing the required flux levels into the material under test. Electromagnets by comparison are excessively bulky and heavy. They do have an advantage in that the magnetic flux levels can be easily adjusted and turned off if necessary for cleaning purposes. Hall effect and coil sensors can be used in this application; however, it has been demonstrated that coil sensors are more sensitive, stable, and reliable.

Magnetic flux leakage is a qualitative, not quantitative, testing tool and is a reliable detector of corrosion on tank floors. However, MFL in this application cannot differentiate between the response from top-side and bottom-side indications. Contrary to what is expected, the flux leakage response from a top-side indication is significantly lower in amplitude than that from an equivalent bottom-side indication.

New Oilfield Tubulars

New oilfield tubulars include oil well casing, tubing, drill pipe, coiled tubing, and line pipe. Manufacturing specifications are governed by the American Petroleum Institute (API) and its equivalent International Organization for Standardization (ISO) standards. Many of the lower grades of these tubulars are inspected by MFL devices in the pipe mill to the sensitivity of various reference indicators, generally EDM notches, and also as customer acceptance tests in storage yards. Tubulars with wall thicknesses greater than aproximately 0.315 in. (8 mm) and high-strength grades are inspected by full-length ultrasound. EDM notches are cut to a depth that is a percentage of the specified wall thickness of the tube and depend on the pipe's grade or the level of inspection demanded by the customer.

Typical MFL systems (Figure 6) consist of a rotating pole yoke (circumferential magnetization) and a surrounding coil (longitudinal magnetization). Appropriate flux-sensitive sensors ride in close proximity to the pipe OD surface. The pipe passes slowly through the rotating electromagnetic yoke, which magnetizes a local section of the pipe wall circumferentially. The pipe is then scanned by arrays of either coils or hall sensors for at least 100% coverage, searching for imperfections with a longitudinal component. The pipe passes into the coil, which then magnetizes it longitudinally, and is scanned while in the coil by sensors that are shaped to the pipe OD for imperfections with a transverse component. The resulting data sets are fed to computers for analysis and then to a recording system. Finally, the tube is partially demagnetized using a coil with a reverse field.

(a)

(b)

Figure 5. Magnetic flux leakage testing of aboveground storage tanks: (a) schematic of bridge; (b) tank floor scanner incorporating MFL test bridge.

Figure 6. Rotating pole and coil system used in casing and tubing MFL inspections.

Internal MFL probes have been replaced by the use of ultrasonic testing (UT) from the outside surface. Various sensor configurations are used to accentuate real signals and suppress noise. Typically, when electric weld tube is inspected, the rotating pole sensors may be connected series opposing.

Because of the longitudinal field demagnetizing effect at the two ends of tubulars, the ends are inspected by magnetic particle testing. This is performed by first inspecting in the residual longitudinal induction from the automated system, then performing a rod slot to rotate the induction into the circular direction and inspecting for longitudinal imperfections in the end areas.

Used Oilfield Tubular Inspections

Used oil-well tubing will suffer internal and external corrosion, slip marks, and internal sucker rod wear. Sucker rods connect to the pump at the bottom of the well and wear both themselves and the inside surface of the tubing, especially in deviated wells. The methods described above for the inspection of new tubes is applicable, but the sensitivity is lowered to accommodate the larger imperfections.

One more modern technique (Figure 7) is to pass the tubing through an inspection head as it is being pulled from the well on a work-over. Here, the pipe is magnetized to saturation longitudinally, and flux leakage is used to roughly determine the depth of internal pitting by connecting surrounding hall sensor arrays in a (+1, −2, +1) configuration axially, which has been found effective for this assessment in real time. Hall effect sensors are used because of the variable pulling speed of the work-over rig.

Used drill pipe stem is inspected by using a traveling head (Figure 8) consisting of a magnetizing coil and encircling sensors in close proximity

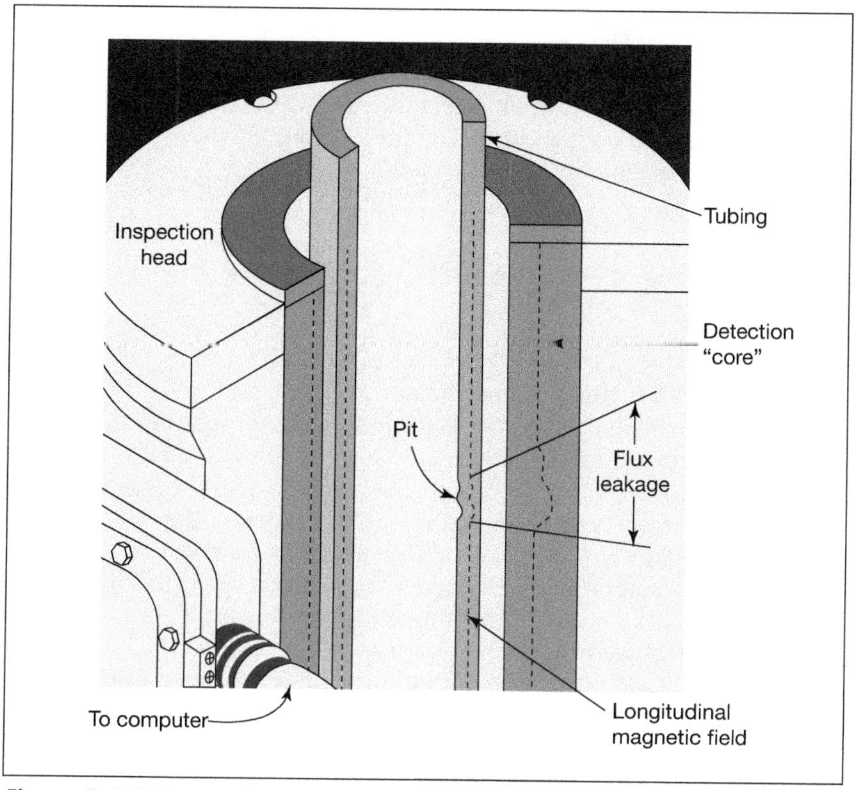

Figure 7. MFL inspection of oil-well tubing being pulled from a well. Hall element sensors are at high liftoff to permit passage of couplings.

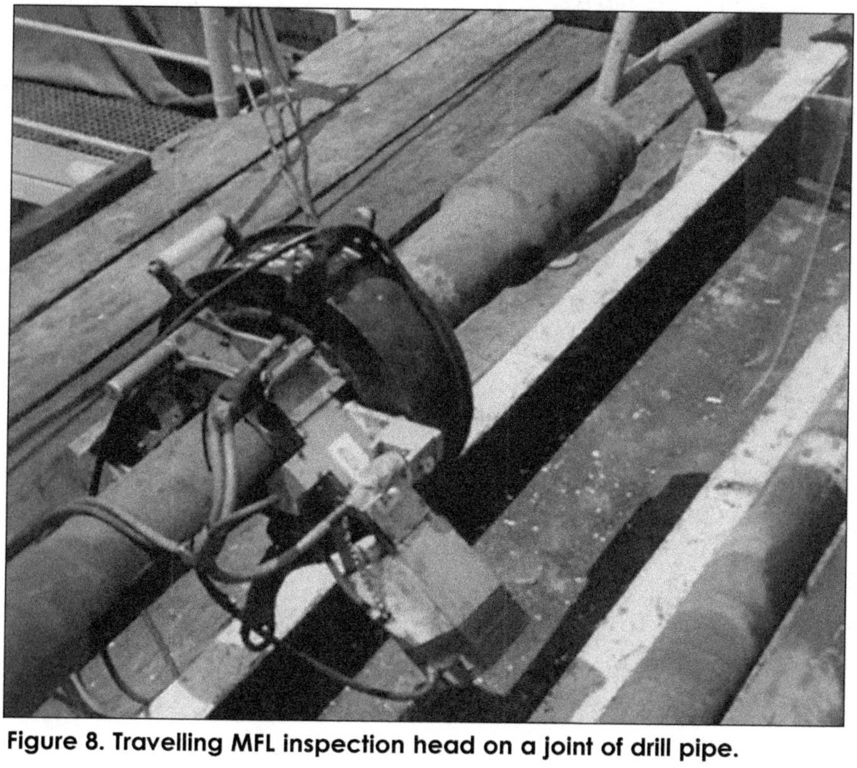

Figure 8. Travelling MFL inspection head on a joint of drill pipe.

to the pipe surface. Flat coil sensors have been developed to accentuate magnetic flux leakage from transverse fatigue cracks but also detect internal pitting and external slip and tong marks. The threads at the pin and box connection ends are inspected in longitudinal induction using wet fluorescent magnetic particle testing.

Unfortunately, with much of oilfield tubular inspection, only the amplitude part of magnetic flux leakage signals is used to initiate prove-up with other methods, such as UT. Other attributes of magnetic flux leakage signals offer greater insight into the source of magnetic flux leakage.

Coiled Oilfield Tubing Inspection

Coiled oilfield tubing is inspected in a similar manner to in-service pipeline inspection but from the outside surface. Longitudinal magnetization from neodymium iron-boron magnets is used, as some of the inspection units are mounted in hazardous areas. The tubing is scanned by surrounding arrays of hall elements, measuring the normal (perpendicular) component of the magnetic flux leakage field from imperfections (B_n). Because of the number of sensors employed, the magnetic flux leakage from each imperfection above a specified threshold is fully recorded and can be viewed and printed as a 3D map of B_n versus axial length versus the circumferential angle. This is helpful in determining the nature of the imperfection. Because coiled tubing is manufactured by electric welding of strips, which themselves have been welded end to end to create strings as long as 30 000 ft (9000 m), the skelp-end welds are also detected since they expel magnetic flux. This is helpful in determining the stretch of the tubing as it elongates during use.

The tangential magnetic field B_t above the surface of the tube is measured by a ring of hall effect sensors that are set back a little from the tubing surface in order to minimize the scanning of magnetic flux leakage (Figure 9). This flux from the magnets has not entered the tube wall but is affected in quantity by that which has. The flux that has entered and left the tube is affected by the poles (P_1, P_2) that appear where the flux enters and leaves, as a demagnetizing effect (H_d) is created within the wall. This demagnetizing effect is affected by the thickness of the tube wall. The result is that B_t is also related in an inverse manner to the wall thickness of a length of the tubing below the sensor. This technique permits wall thickness to be measured so long as it does not change rapidly with length and is effective for pipe covered in rust or oil, where UT cannot be used for wall thickness measurement.

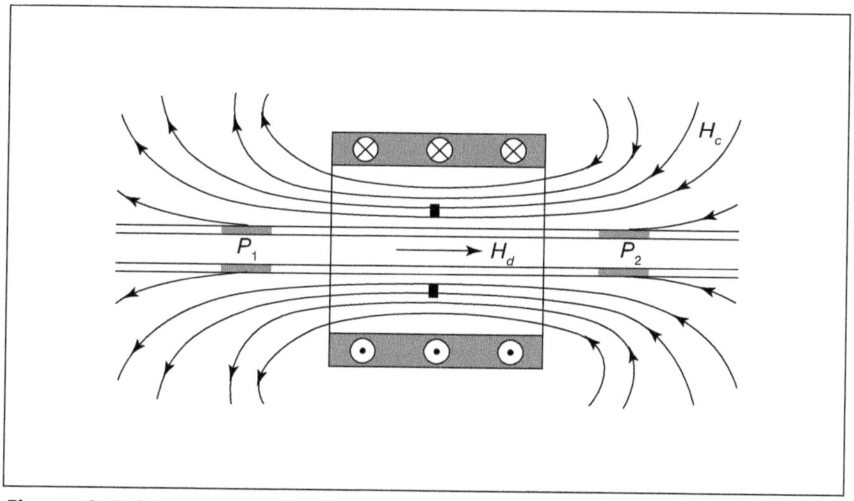

Figure 9. Tubing magnetization scheme for magnetic wall thickness measurement with fields H_c and H_d shown.

›16

Electromagnetic Testing Standards

Introduction

Two types of standards are discussed in this chapter:

- Reference standards: a physical object with known material characteristics used as a basis for comparison or standardization. For example, in tube testing, a tube with artificial discontinuities is used for establishing the test sensitivity setting and for periodically checking and adjusting the sensitivity setting, as required.
- Industry specifications and standards: documents to control and govern practices in an industry or application, applied on a national or international basis and usually produced by consensus.

Reference Standards

An important requirement for successful ET is the use of an accurate reference standard for equipment standardization. The reference standard is used to adjust the electromagnetic equipment's sensitivity to various test object parameters (cracks, surface roughness, conductivity/permeability variations, and other material conditions).

Success in ET depends on the proper use of available reference standards. The development and use of electromagnetic reference standards requires a thorough understanding of the test to be performed. Reference standard considerations should include:

- Material tested.
- Size and shape.
- Discontinuities of interest.
- Means of producing artificial discontinuities.
- Nonrelevant indications that might be encountered, such as material property variations.
- Instrument capabilities and limitations.
- Criteria for relevant indications.

In view of many types of artificial discontinuities that can be produced, an understanding of their relationship to the discontinuities of interest is critical. For example, if transverse discontinuities in a tube are of interest, it might seem obvious to use a transverse notch. However, that notch could be either straight, which is much easier to fabricate, or curved to match the radius of the tube. Figure 1 illustrates this configuration. A similar case is that of cracks in a fastener hole, which

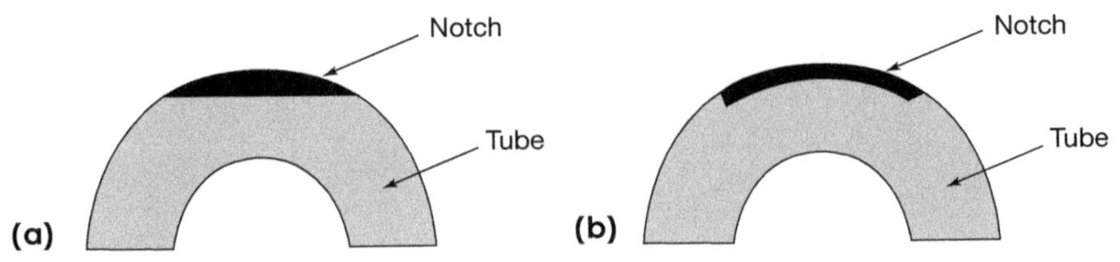

Figure 1. Transverse notches in tube: (a) straight; (b) curved.

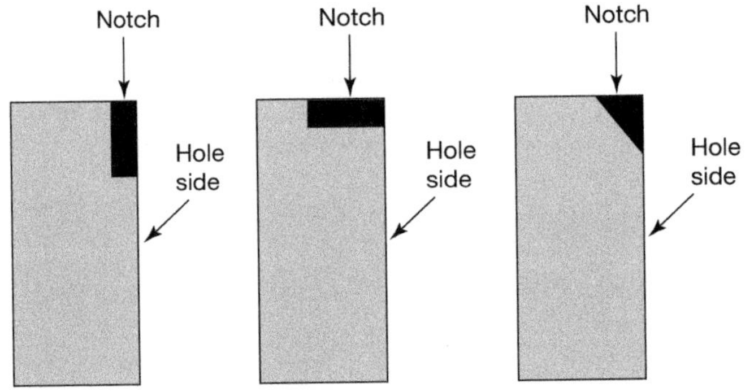

Figure 2. Notches to simulate cracks as might be found in fastener holes.

can be simulated by several notches, as shown in Figure 2, all of which produce different signals.

Reference Standards

Conductivity Reference Standards

In electromagnetic testing, frequent use is made of measurements of conductivity based on the International Annealed Copper Standard (IACS). Two standard metal blocks are supplied with conductivity measuring instruments. One block represents a high level of conductivity, while the other represents a low level of conductivity. The percentage value in IACS is stamped on the blocks, as illustrated in Figure 3.

Conductivity reference standards should be tested with a relatively small coil to determine the uniformity of electrical conductivity over the

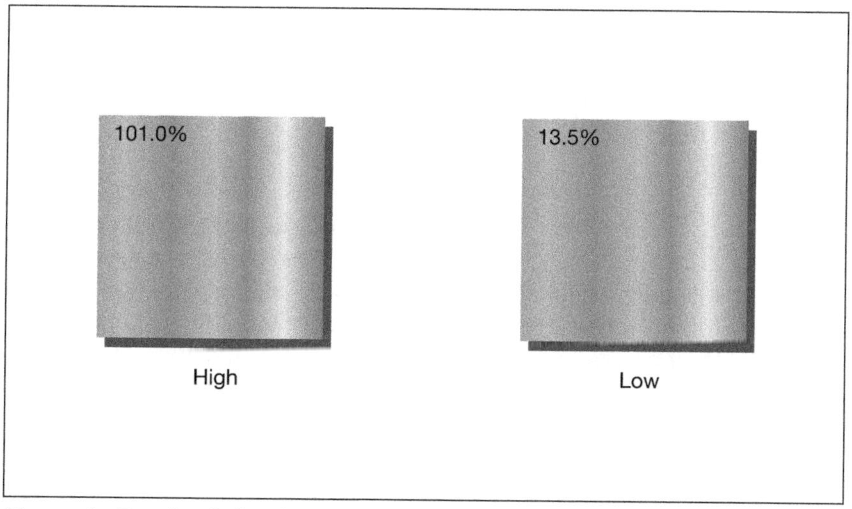

Figure 3. Conductivity standards.

surface of the standard. Both the front and the back surface should be tested for conductivity differences.

Coating Thickness Reference Standards

Coating thickness reference standards of uniform thickness are available in two types:

- Foils or shims of known thicknesses laid on an appropriate substrate.
- Actual coatings affixed to prepared substrates as supplied or recommended by the instrument manufacturer or standardizing organization.

Table 1 lists standard reference materials from the National Institute of Standards and Technology (NIST) for the standardization of instruments used in measurement of organic and nonmagnetic inorganic coatings over steel. Each 1.8 × 1.8 in. (4.5 × 4.5 cm) block consists of fine-grained copper electrodeposited on low carbon steel substrate.

Table 1. NIST reference materials

Material number	Coating thickness	
	µm	(in. $\times 10^{-3}$)
1358a	80, 255, 1000	(3.1, 9.8, 39)
1359b	48, 140, 505, 800	(2.0, 5.5, 20, 32)
1361b	6, 12, 25, 48	(0.2, 0.5, 1.0, 2.0)
1362b	40, 80, 140, 205	(1.6, 3.1, 5.5, 7.9)
1363b	255, 385, 505, 635	(9.8, 16, 20, 26)

Discontinuity Reference Standards

A discontinuity reference standard should duplicate the test situation for material type and geometry. In addition, the standard should include discontinuities that establish the maximum acceptable discontinuity and the sensitivity of the test system. Discontinuity reference standards fall under two types:

- Natural discontinuity reference standards.
- Artificial discontinuity reference standards.

Natural Discontinuity Reference Standards

Natural discontinuity standards consist of duplicates of the test piece configuration that contain natural discontinuities of a known size and shape. At least one of the discontinuities should be at the limit of acceptability. Natural discontinuities can be developed or accumulated. For example, cracks can be developed by submitting a material to cyclic stresses until a natural fatigue crack is generated. This can then be machined to produce a surface or hole crack of known depth. Figure 4 illustrates an example of how a natural fatigue crack is developed.

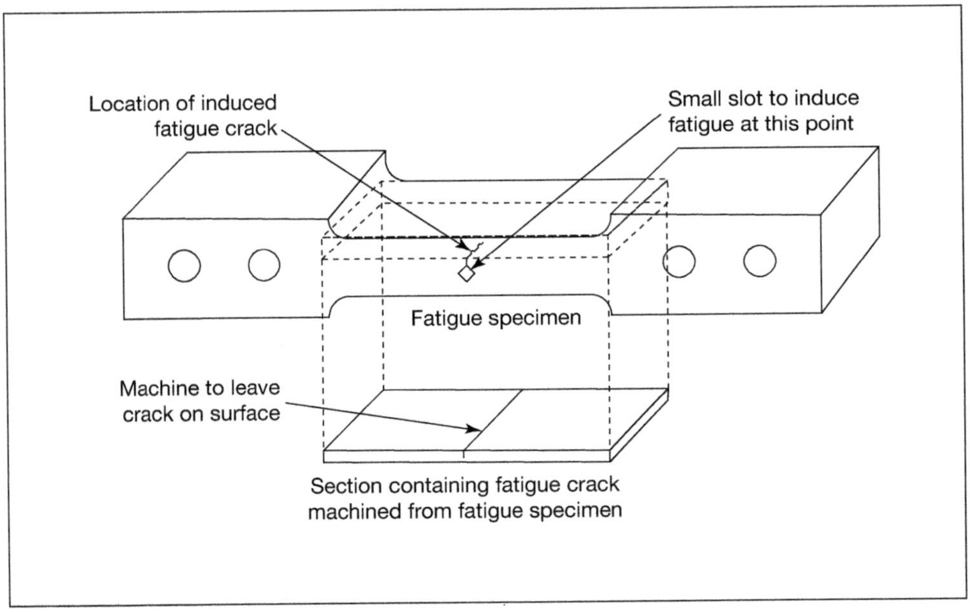

Figure 4. Development of a natural fatigue crack.

Artificial Discontinuity Reference Standards

Artificial discontinuities may be machined into a duplicate of the test object configuration. Several samples may have to be run through the test system to find one that does not produce any appreciable indications of natural discontinuities from the test piece prior to machining. Standard reference discontinuities that are pertinent to the required specification are then fabricated into the sample.

Types of standard reference discontinuities that can be used to simulate test object discontinuities include longitudinal notches, circumferential notches, drilled holes, flat-bottom holes, diameter steps, and geometry variations (Figure 5).

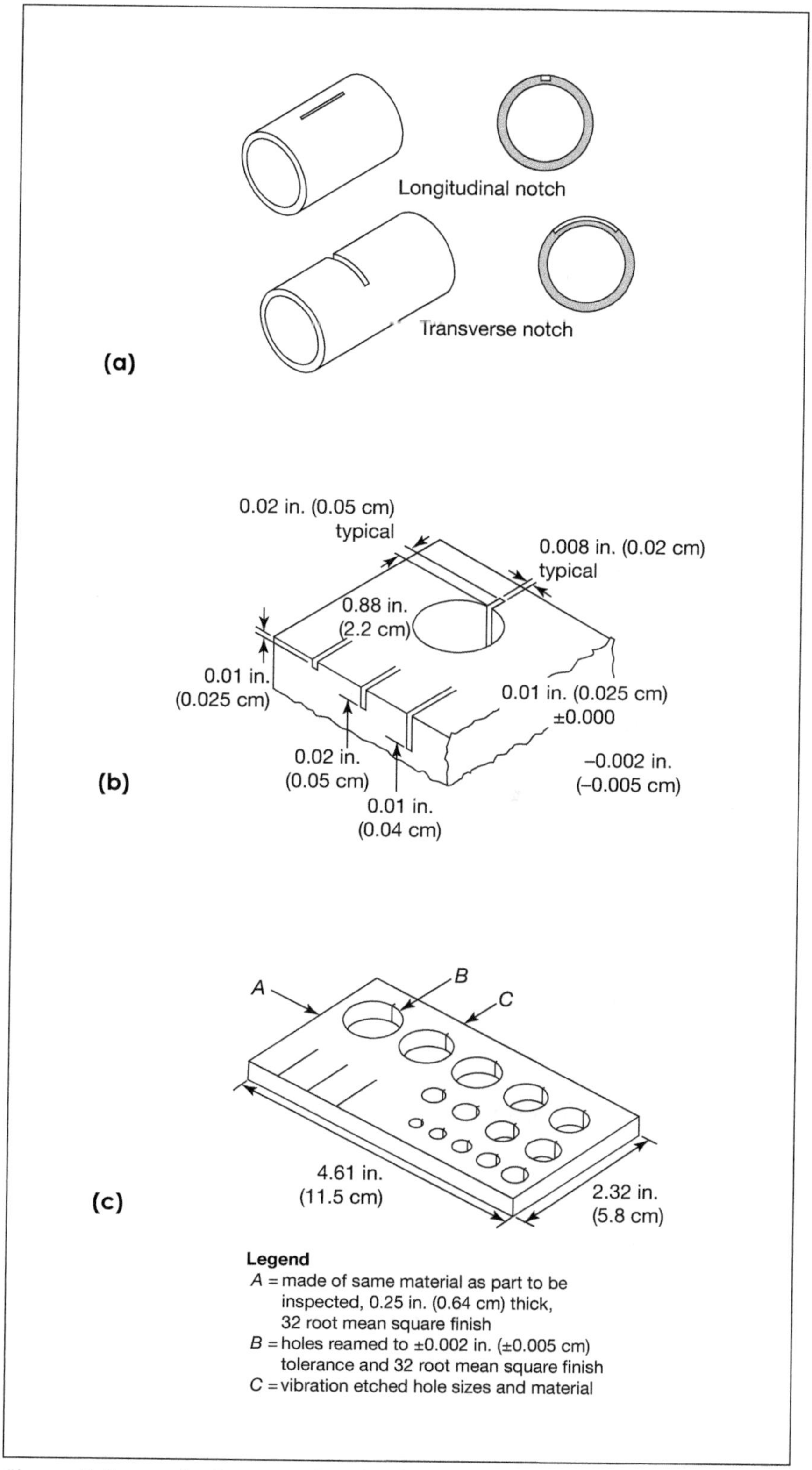

Figure 5. Type of reference standards: (a) notched tubes; (b) reference block; (c) block with graduated holes.

When external comparison techniques are used, the reference standard must be free of discontinuities. The standard must be representative of imperfections that may be found in the test object.

Liftoff Reference Standards

Liftoff reference standards are easily constructed by the application of known thicknesses of a nonconductive material to a sample of the test object. Paper, polyethylene terephthalate, and cellophane are examples of nonconductive materials often used. When it is required to measure the thickness of nonconductive coating, the standard does not have to be coated with the same nonconductive coating material since, with ET, one nonconductive coating is exactly like any other. The firm requirement is that the thickness of the coating used on the standard be known.

If measuring thickness of a nonconductive coating over a conductive test object, liftoff standards need to be constructed that represent both the maximum and the minimum acceptable standard.

Sorting Reference Standards

When sorting with the absolute encircling coil technique, a known acceptable reference standard and a known unacceptable standard are required. When using the comparative encircling coil technique, usually two known acceptable specimens of the test object and one unknown unacceptable specimen are required. For a three-way sort, it is best to have three reference standards, including either two for the high and low limits of acceptability for one group, or one each for two unacceptable groups. The third reference standard represents the acceptable lot of material.

Although electromagnetic sorters can be useful for limited applications, sorting steels is better done via spectroscopic equipment. A change in product diameter, for example, can simulate a change in composition to electromagnetic sorters.

Procedures and Specifications

Procedures and specifications are produced to provide a means of controlling product or service quality. Written instructions that can be used as guidance to companies or individuals to a desired end result and that are acceptable to industry form the basis for these kinds of documents.

In nondestructive testing, a *specification* is often written by a commercial organization, usually one of the primary parties in the purchasing agreement. A specification is product specific and may be considered a tailored form of a standard. A specification can require more stringent limits than a related standard's limit. In practice, a specification provides a clearly organized list of testing parameters (a specially written procedure) that describes the technique for locating and categorizing discontinuities in a specific test object. A typical specification includes acceptance criteria and is required by the designer, buyer, manufacturer, and end users of the test object it covers.

A *procedure* refers to a set of brief generalized guidelines that show the technician how to perform an accurate test for a given contract. A procedure often includes details about the in-house setup. Specifications are written to eliminate the variable characteristics of human operators and system designs, to produce an accurate result regardless of who is performing the electromagnetic test. Specifications must be written with full knowledge of the electromagnetic test technique, its sensitivity, test object design, and its material characteristics. In addition, the specification must provide the technician with knowledge of discontinuities critical to the test object's service life.

Testing specifications are working documents that instruct how to locate discontinuities in a specific test object. As such, specifications are in regular need of review and revision.

Standards and Industry Specifications

Several publications are available to guide or instruct ET procedures. Some of the most frequently used references include the following.

ASTM International

ASTM International (formerly, the American Society for Testing Materials) publishes several standards pertaining to electromagnetic testing techniques. These standards are numbered, for example, *E268-81*. *E268* refers to the standard, and "81" refers to the year of origin or the last revision. ASTM standards that pertain to electromagnetic techniques include:

- *E309-16: Standard Practice for Eddy Current Examination of Steel Tubular Products using Magnetic Saturation*
- *E376-11: Standard Practice for Measuring Coating Thickness by Magnetic-Field or Eddy-Current (Electromagnetic) Testing Methods*
- *E426-16: Standard Practice for Electromagnetic (Eddy Current) Examination of Seamless and Welded Tubular Products, Titanium, Austenitic Stainless Steel and Similar Alloys*
- *E566-14: Standard Practice for Electromagnetic (Eddy Current) Sorting of Ferrous Metals*
- *E570-15el: Standard Practice for Flux Leakage Examination of Ferromagnetic Steel Tubular Products*
- *E703-14: Standard Practice for Electromagnetic (Eddy Current) Sorting of Nonferrous Metal*

American Society of Mechanical Engineers

In 1911, the American Society of Mechanical Engineers (ASME) set up a committee to establish rules for safety for design, fabrication, and testing of boilers and pressure vessels. These rules have become known throughout industry as ASME code.

The *ASME Boiler & Pressure Vessel Code* is divided into 11 sections. Section V, "Nondestructive Examination," is divided into two subsections, A and B. Subsection A deals with nondestructive methods of examination. Article 8 covers Eddy Current Examination of Tubular Products. Subsection B contains documents adopted by Section V. Electromagnetic testing standards are described in Article 26. In this case, *ASTM E215-16: Standard Practice for Standardizing Equipment for Electromagnetic Examination of Seamless Aluminum-Alloy Tube*, has been adopted by ASME and reassigned the designation SE 215.

Military Standards

The US military uses the military standard document to control testing and materials. Standard procedures are provided by a series of MIL-STD-XXXX documents. Special requirements are specified by the military specification system. The military standard usually contains several parts and is very descriptive. These parts normally include scope, applicable documents, definitions, general requirements, detail requirements, and notes. With recent downscaling in military specification efforts, many of these specifications have been incorporated into the Society of Automotive Engineers' *AMS 2644* and industry is in transition towards the use of this specification.

American Petroleum Institute

The American Petroleum Institute (API) sets ANSI-approved standards for the oil and gas industry. (ANSI stands for the American National Standards Institute.) Those that are specific to MFL of tubulars, which are generally either EDM notches or drilled holes, are as follows:

- *API Spec 5CT: Specification for Casing and Tubing*
- *API Spec 5L: Specification for Line Pipe*
- *API Spec 5ST: Specification for Coiled Tubing*
- *API Spec 5LCP: Specification for Coiled Line Pipe*
- *API Spec 5D: Specification for Drill Pipe*
- *API RP 5A5: Field Inspection of New Casing, Tubing, and Plain-End Drill Pipe*

Appendix A: Formulas

Magnetic Permeability

The ease with which a material can be magnetized:

(Eq. 1) $$\mu = \frac{B}{H}$$

where

μ is permeability (H/m)
B is flux density (gauss or tesla)
H is magnetizing force (oersted or ampere/meter)

Magnetic Permeability in Free Space (Air)

Magnetic permeability, μ (mu), is a combination of terms.
For nonferromagnetic materials:

(Eq. 2) $$\mu_0 = 4\pi \times 10^{-7}\,\text{H/m}$$

Relative Permeability (Dimensionless)

For ferromagnetic materials:

(Eq. 3) $$\mu_r = \frac{\mu}{\mu_0}$$

where

μ_r is relative permeability (dimensionless)
μ_0 is $4\pi \times 10^{-7}$ H/m

Resistance of a Coil

Opposition to transmission of electrical current:

(Eq. 4a) $$\text{resistance} = \frac{(\text{specific resistance}) \times (\text{length})}{\text{area}}$$

or

(Eq. 4b) $$R = 2.54 \times \frac{\rho \times L}{A}$$

where
resistance is ohms
specific resistance is ohms per circular mil-foot
area is circular mils

Resistivity

Ability of material to resist electrical current:

(Eq. 5) $$\rho = \frac{1}{\sigma}$$

where
ρ is resistivity ($\Omega \cdot$cm)
σ is conductivity (siemens/meter)

Conductivity

Ability of material to transmit electric current:

(Eq. 6a) $$\sigma = \frac{1}{\rho}$$

where
σ is conductivity (siemens/meter)
ρ is resistivity ($\Omega \cdot$cm)

Expressed in %IACS:

(Eq. 6b)
$$\%\text{IACS} = \frac{172.41}{\text{resistivity}(\rho), \text{in } \mu\Omega\text{cm}}$$
$$\text{resistivity}(\rho), \text{in } \mu\Omega\text{cm} = \frac{172.41}{\%\text{IACS}}$$

Ohm's Law

(Eq. 7) $$E = I \times R$$

where

E is electromotive force (V)
I is electric current (A)
R is resistance (Ω)

Power

(Eq. 8a) $$P = I \times E$$

or

(Eq. 8b) $$P = I^2 R$$

where

P is power (W)
I is current (A)
E is voltage (V)
R is resistance (Ω)

Coil Inductive Reactance

Opposition to changes in alternating current flow through a coil:

(Eq. 9a) $$X_L = \omega_L$$

or

(Eq. 9b) $$X_L = 2\pi f L = \omega_L$$

where

X_L is inductive reactance (Ω)
π is 3.1416
f is frequency (Hz)
L is inductance (H)
ω is angular frequency (rad/s)

Air Coil Induction

(Eq. 10a) $$L = \frac{0.8(rN)^2}{6r + 9l + 10b}$$

where
- L is self-inductance (μH)
- N is number of turns in coil
- r is mean radius (in.)
- l is length of coil (in.)
- b is coil depth or thickness (in.)

Coil induction can also be calculated using this formula:

(Eq. 10b) $$L = \frac{N^2 A \times \mu_r \mu_0}{l}$$

where
- L is inductance (H)
- N is number of turns in coil
- μ_r is relative permeability (dimensionless)
- l is length (m)

Capacitive Reactance

Opposition to changes in alternating voltage across a capacitor:

(Eq. 11) $$X_C = \frac{1}{2\pi f C}$$

where
- X_c is capacitive reactance (Ω)
- f is frequency (Hz)
- C is capacitance (farads)

Coil Impedance

The total opposition to alternating current flow:

(Eq. 12) $$Z = \sqrt{{X_L}^2 + R^2}$$

where
- Z is coil impedance (Ω)
- X_L is coil inductive reactance (Ω)
- R is coil resistance (Ω)

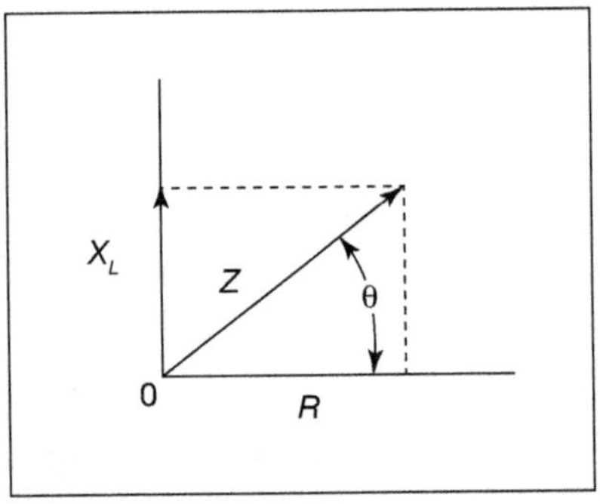

Figure 1. Impedance-plane diagram.

Coil Impedance Phase Angle

Phase is a time delay expressed in terms of 360° alternating cycle:

(Eq. 13a)
$$\tan\theta = \frac{X_L}{R}$$

where

θ is phase angle (°)
X_L is coil inductive reactance (Ω)
R is coil resistance (Ω)

Thus:

(Eq. 13b)
$$\theta = \arctan\frac{X_L}{R}$$

Phase lag can be looked at as how long (in phase/time) it takes energy to reach a certain point in the test material. This answer directly relates to the displayed phase angle information we see between two events expressed as responses on the tester screen.

Phase lag in radians:

(Eq. 13c)
$$\theta = \frac{x}{\delta}$$

where

δ is depth of penetration (in.)

Phase lag in degrees:

(Eq. 13d)
$$\theta = \frac{x}{\delta}57.3^{\circ}$$

θ is radians lagging (°)
x is depth of interest (in.)
δ is depth of penetration (in.)

Note: Metric equivalents could be used if applied to both unknowns using the same scaling factor.

Coil Merit

This term is used to describe coil efficiency Q or merit of the coil. The higher the Q or merit of a coil, the more efficiently the coil performs as an inductor.

(Eq. 14)
$$Q = \frac{X_L}{R}$$

where
Q is merit of the coil
X_L is inductive reactance (Ω)
R is resistance (Ω)

Note: The addition of permeable core materials in certain designs dramatically improves the Q factor. Permeable cores are usually constructed of high-permeability powdered iron.

Standard Depth of Penetration (Eddy Current Testing)

(Eq. 15a)
$$\delta = \frac{1}{\sqrt{\pi f \mu \sigma}}$$

where
δ is standard depth of penetration (m)
π is 3.1416
f is frequency (Hz)
μ is magnetic permeability (H/m)
σ is electrical conductivity (mhos/m)

The standard depth of penetration also can be calculated using the following formula:

(Eq. 15b)
$$\delta = K\sqrt{\frac{\rho}{f\mu_{rel}}}$$

where
δ is standard depth of penetration (m)
K is 1.98 (for in.) or 50 (for mm scale)
ρ is resistivity (μΩ·cm)
f is frequency (Hz)
μ_{rel} is 1 (for nonferromagnetic materials)

Frequency Calculation for Inspection of Heat Exchanger Tubes (Eddy Current)

Sizing frequency:

(Eq. 16)

$$f_0 = \frac{10\rho}{t^2}$$

where

f_0 is optimum frequency (Hz)
ρ is electrical resistivity (μΩ·cm)
t is tube wall thickness (in.)

Note: For f_{90} (detection frequency), the ET instrument's operating frequency could be selected as half optimum frequency. A phase spread of approximately 90° between the signal response from the four 20% flat-bottom holes and the 360° ID groove using a typical ASME standard may be achieved.

For remote field testing of ferromagnetic tubing, there is no valid calculation for an exact operating frequency. The permeability values for a given tube in a given heat exchanger could be very different from the reference standards used prior to conducting a field test. The only way to validate that the operating test frequency is correct is to review the test data and make changes, as required.

As a general rule, heavier wall tubing (16 gauge or thicker) with higher permeability alloys (carbon steel) will probably be inspected in the range of 400 Hz or less. As tubing wall thickness decreases and/or alloys have decreasing permeability, the test frequency will rise above 400 Hz. For some nickel-based alloys, the frequencies may rise into the lower kilohertz range (1 to 20 kHz).

(Eq. 17)

$$\eta = \frac{d^2}{D^2}$$

where

η = fill factor

Inside diameter coils:

d = coil windings outside diameter
D = tube internal diameter

Outside diameter coils:

D = coil windings inside diameter
d = test object outside diameter

Appendix B: Tables

Table 1. Standard depth of penetration for eddy current testing of nonferromagnetic alloys ($\mu_r = 1$)

Depth	Amplitude compared to surface value	Phase lag
0 (surface)	100%	0 rad = 0°
1δ	1/ε* = 37%	1 rad = 57.3°
2δ	$1/\varepsilon^2$ = 14%	1 rad = 114.6°
3δ	$1/\varepsilon^3$ = 5%	1 rad = 171°
At a depth of one delta (δ) the eddy current strength has decreased to 37% of its original surface value. The theoretical phase lag at that point is 57°. *ε (epsilon) = 2.718		

Table 2. Electromagnetic quantities and SI units

Symbol	Quantity/description	Unit
I	Current	ampere (A)
E	Electrical potential	volt (V)
P	Power	watt (W)
R	Resistance	ohm (Ω)
Z	Impedance	ohm (Ω)
X_L	Inductive reactance	ohm (Ω)
X_C	Capacitive reactance	ohm (Ω)
C	Capacitance	farad (F)
T	Amount of time per cycle	second (s)
π	Pi	3.1416
σ (sigma)	Conductivity	%IACS (siemens per meter)
δ (delta)	Standard depth of penetration	metric or English
B	Flux density	tesla (T); gauss (G);
μ (mu)	Magnetic permeability	weber (Wb)/amp-meter (A-m)
H	Magnetizing force	oersted (Oe) or amp/meter (A/m)
Ø	Phase angle lag	radians (degrees)
D	Coil diameter	metric or English
f	Frequency	hertz (Hz)
N	Number of turns (windings)	dimensionless
ρ (rho)	Electrical resistivity	μΩ·cm
L	Self-inductance	henry (H)
J	Current density	amp per square meter (A/m^2)
η (eta)	Fill factor	dimensionless
A	Cross-sectional area	metric or English
t	Thickness	metric or English

References

The following publications are training references for further study as recommended in *ANSI/ASNT CP-105: ASNT Standard Topical Outlines for Qualification of Nondestructive Testing Personnel* (2016).

Annual Book of ASTM Standards, Vol. 03.03, *Nondestructive Testing*, ASTM International, Philadelphia, PA, latest edition.†

ASNT Level III Study Guide: Electromagnetic Testing, American Society for Nondestructive Testing, Inc., Columbus, OH, latest edition.*

ASNT Questions & Answers Book: Electromagnetic Testing, American Society for Nondestructive Testing, Inc., Columbus, OH, latest edition.*

ASTM E 690-10 Standard Practice for In Situ Electromagnetic (Eddy-Current) Examination of Nonmagnetic Heat Exchanger Tubes, ASTM International, West Conshohocken, PA, 2010.†

Beissner, R.E., G.A. Matzkanin, and C.M. Teller, *NTIAC-80-1, NDE Applications of Magnetic Leakage Field Methods*, January 1980.

Bray, D.E., and D. McBride, 1992, *Nondestructive Testing Techniques*, John Wiley & Sons, New York, NY.

Cecco, V.S., G. Van Drunen, and F.L. Sharp, 1987, *Eddy Current Testing*, US Edition, Nichols Publishing, Inc., Columbia, MD, (AECL-7523).

Kilgore, R.J., and S. Ramchandran, 1989, "Remote-field Eddy Current Testing of Small-diameter Carbon Steel Tubes," *Materials Evaluation*, Vol. 47, No. 1, pp. 32–36.*

Libby, H.L., 1979, *Introduction to Electromagnetic Nondestructive Test Methods*, Robert E. Krieger Publishing Co., Huntington, NY.

Mackintosh, D.D., D.L. Atherton, and P.A. Puhach, 1993, "Through-transmission Equations for Remote-field Eddy Current Inspection of Small-bore Ferromagnetic Tubes," *Materials Evaluation*, Vol. 51, No. 6, pp. 744–748.*

Mackintosh, D.D., D.L. Atherton, T.R. Schmidt, and D.E. Russell, 1996, "Remote Field Eddy Current for Examination of Ferromagnetic Tubes," *Materials Evaluation*, Vol. 54, No. 6, pp. 652–657.*

Materials and Processes for NDT Technology, 2016, second edition, American Society for Nondestructive Testing, Inc., Columbus, OH.*

Mordfin, L., *Handbook of Reference Data for Nondestructive Testing*, 2002, second edition, ASTM International, West Conshohocken, PA.†

Sadek, H., 2006, *Electromagnetic Testing Classroom Training Book*, (PTP Series), first edition, American Society for Nondestructive Testing, Inc., Columbus, OH.*

Schmidt, T.R., 1989, "History of the Remote Field Eddy Current Inspection Technique," *Materials Evaluation*, Vol. 47, No. 1, pp. 14–22.*

Schmidt, T.R., 1984, "The Remote Field Eddy Current Technique," *Materials Evaluation*, Vol. 42, No. 2, pp. 223–230.*

Schmidt, T.R., D.L. Atherton, and S. Sullivan, 1989, "The Remote-field Transition Zone," *Materials Evaluation*, Vol. 47, No. 9, pp. 969–979.*

Smith, H., and D.D. Mackintosh, "Remote Field Eddy Current Examination of Boiler Tubes," *Proceedings of EPRI Topical Workshop on Electromagnetic NDE Application in the Electric Power Industry*, Charlotte, NC, 21–23 August 1995, pp. 1–17.

Sollish, D.B., "Field Experience in Boiler Examinations using Remote Field Eddy Current," Corrosion 91: NACE Annual Conference and Corrosion Show, Cincinnati, OH, 11–15 March 1991, pp. 224/1–224/8.

Sullivan, S., and D.L. Atherton, 1989, "Analysis of the Remote Field Eddy Current Effect in Non-magnetic Tubes," *Materials Evaluation,* Vol. 47, No. 1, pp. 80–86.*

Taylor, J.L., ed., 1988, *Basic Metallurgy for Non-Destructive Testing*, revised edition, W.H. Houldershaw, Ltd. (British Institute of Non-Destructive Testing), Essex, England.*

Udpa, S., tech. ed., and P.O. Moore, ed., 2004, *Nondestructive Testing Handbook*, third edition: Vol. 5, *Electromagnetic Testing*, American Society for Nondestructive Testing, Inc., Columbus, OH.*

---

*Available from the American Society for Nondestructive Testing, Inc.
†Available from ASTM International.

Figure Sources

All figures derive from sources published by The American Society for Nondestructive Testing, Inc., except for the following:

Chapter 2

Figure 7 - NDT International, MFE
Figure 8 - Queen's University, AMG

Chapter 6

Figures 5, 6, 7 - Foerster
Figure 8b - C. Dodd, ORNL

Chapter 12

Figures 1, 2a, 2b, 3, 4, 5, 6, 7, 8, 9, 10, 11, 12, 13, 14, 15, 16, 17, 18, 19, 20, 21, 22, 23, 24, 25 - Used with permission of Technical Software Consultants, Ltd., UK

Chapter 15

Figures 6, 7, 8, 9 - R.K. Stanley

Glossary

Absolute coil: Coil that responds to the electromagnetic properties of that region of the test part within the magnetic field of the coil, without comparison to the response of a second coil at a different location on the same or similar material.

Absolute measurement: (1) Measurement made with an absolute coil. (2) Measurement of a property without reference to another measurement of that property.

Acceptance criterion: Benchmark against which test results are to be compared for purposes of establishing the functional acceptability of a part or system being examined.

Acceptance level: Measured value or values at, above, or below which test objects are acceptable in contrast to rejection level.

Acceptance limit: Test signal value used in electromagnetic testing, establishing the group to which a material under test belongs.

Acceptance standard: Specimen, similar to the product to be tested, containing natural or artificial discontinuities that are well defined and similar in size or extent to the maximum acceptable in the product. Acceptance standards are available also for material properties such as conductivity and hardness.

Alternating current: Electrical current that reverses its direction at regular intervals.

Alternating current field: Varying magnetic field produced around a conductor by alternating current flowing in the conductor.

Alternating current magnetization: Magnetization by a magnetic field generated when alternating current is flowing.

Ampere (A): SI unit of electric current.

Amplitude response: Property of a test system whereby the amplitude of the detected signal is measured without regard to phase. See also *Phase analysis*.

Analog-to-digital converter: Circuit whose input is information in analog form and whose output is essentially the same information in digital form.

Anisotropy: Having properties that differ according to the direction of measurement.

Artificial discontinuity: See *Discontinuity, artificial*.

Artificial discontinuity standard: See *Acceptance standard*.

ASNT: American Society for Nondestructive Testing.

Attenuation: Decrease in signal amplitude over distance, often called *loss*; can be expressed in decibels or as a scalar ratio of the input magnitude to the output magnitude.

Automated system: Acting mechanism that performs required tasks at a determined time and in a fixed sequence in response to certain conditions.

Bandpass filter: Frequency filter that has a single transmission band between two cutoff frequencies, neither of the cutoff frequencies being zero or infinity.

Bandwidth: Difference between the cutoff frequencies of a bandpass filter.

Calibration, instrument: Comparison or adjustment of an instrument to a known reference(s), often traceable to the National Institute of Standards and Technology. See also *Standardization, instrument*.

Central conductor: Electric conductor passed through the opening in a part with an aperture, or through a hole in a test object, for the purpose of creating a circular magnetic field in the object.

Certification: With respect to nondestructive testing personnel, process of providing written testimony that an individual is qualified. See also *Certified* and *Qualified*.

Certified: With respect to nondestructive testing personnel, having written testimony of qualification. See also *Certification* and *Qualification*.

Circular magnetization: Magnetization in an object resulting from current passed longitudinally through the object itself or through an inserted central conductor.

Circumferential coil: See *Encircling coil*.

Coil: One or more loops of a conducting material; a single coil may be an exciter and induce currents in the material or it may be a detector or both simultaneously.

Coil clearance: See *Liftoff*.

Coil spacing: In electromagnetic testing, the axial distance between two encircling or inside coils of a differential or remote field test system.

Conductance: Transmission of electric current through material. Measured in siemens (S). Inversely related to resistance *R* (ohm).

Conductivity: Ability of material to transmit electric current. Measured in siemens per meter. Inversely related to resistivity, ρ.

Contact head: Electrode assembly used to clamp and support an object to facilitate passage of electric current through the object for circular magnetization.

Coupled: (1) Of two electric circuits, having an impedance in common so that a current in one causes a voltage in the other. (2) Of two coils, sharing parts of their magnetic flux paths.

Coupling: Percentage of magnetic flux from a primary circuit that links a secondary circuit; effectiveness of a coil in inducing eddy currents in the test object.

Current flow technique: Magnetizing by passing current through an object using prods or contact heads. The current may be alternating current or rectified alternating current.

Current induction technique: Magnetization in which a circulating current is induced in a ring component by a fluctuating magnetic field.

Cycle: Single period of a waveform or other variable. See *Period*.

Defect: Discontinuity whose size, shape, orientation, or location make it detrimental to the useful service of its host object or which exceeds the accept/reject criteria of an applicable specification. Note that some discontinuities may not exceed specifications and are therefore not defects. Compare *Discontinuity* and *Indication*.

Demodulation: Process wherein a carrier frequency modulated with a signal of lower frequency than the carrier frequency is converted to a close representation of the original modulating signal. See *Modulation*.

Depth of penetration: See *Skin effect* and *Standard depth of penetration*.

Differential amplifier: Amplifier whose output signal is proportional to the algebraic difference between two input signals.

Differential coils: Two or more physically adjacent and mutually coupled coils connected in series opposition such that an imbalance between them, causing a signal, will be produced only when the electromagnetic conditions are different in the regions beneath two of the coils. In contrast, comparator coils are not adjacent or mutually coupled.

Differential measurement: In electromagnetic testing, the measurement of system imbalance by using differential coils, in contrast to absolute and comparative measurements.

Differentiated signal: In electromagnetic testing, an output signal proportional to the input signal's rate of change.

Direct current: Electric current flowing continually in one direction through a conductor without variation in amplitude.

Direct current field: Active magnetic field produced by direct current flowing in a conductor or coil.

Discontinuity: Interruption in the physical structure or configuration of a test object. After nondestructive testing, unintentional discontinuities interpreted as detrimental to the serviceability of the host object may be called *flaws* or *defects*. Compare *Defect* and *Indication*.

Discontinuity, artificial: Reference discontinuity such as a hole, indentation, crack, groove, or notch introduced into a reference standard to provide accurately reproducible indications for determining sensitivity levels.

Domain: Any of numerous contiguous regions in a ferromagnetic material in which the direction of spontaneous magnetization is uniform and different from that in neighboring regions.

Eddy current: Electrical current induced in a conductor by a time-varying magnetic field.

Eddy current testing: Electromagnetic testing technique in which eddy current flow is induced in the test object. Changes in the flow caused by variations in the specimen are reflected into a nearby coil, coils, hall effect device, or other magnetic flux sensor for subsequent analysis by suitable instrumentation and techniques.

Edge effect: In electromagnetic testing, the disturbance of the magnetic field and eddy currents because of the proximity of an abrupt change in geometry, such as an edge of the test object. Sometimes called *end effect*. The effect generally results in the masking of discontinuities within the affected region.

Effective depth of penetration: In electromagnetic testing, the minimum depth beyond which a test system can no longer practically detect a further increase in specimen thickness.

Electric field: Vector field of either the electric field intensity (volt per meter) or of the electric flux density (coulomb per meter squared).

Electromagnet: Ferromagnetic core surrounded by a coil of wire that temporarily becomes a magnet when an electric current flows through the wire.

Encircling coil: In electromagnetic testing, a coil or coil assembly that surrounds the test object. Such a coil is also called an *annular coil, circumferential coil,* or *feed-through coil.*

End effect: See *Edge effect.*

ET: Electromagnetic testing.

Evaluation: Review following interpretation of indications, to determine whether they meet specified acceptance criteria.

Excitation coil: Coil that carries the excitation current. Also called *primary coil* or *winding.*

External discontinuities: Discontinuities on the outside or exposed surface of a test object.

False indication: Test indication that could be interpreted as originating from a discontinuity but that actually originates where no discontinuity exists in the test object. Distinct from nonrelevant indication. Compare *Defect.*

Ferromagnetic material: Material such as iron, nickel, or cobalt whose relative permeability is considerably greater than unity and depends on the magnetizing force and often exhibits hysteresis. Materials that are most strongly affected by magnetism are called *ferromagnetic.*

Fill factor: For encircling coil electromagnetic testing, the ratio of the cross-sectional area of the test object to the effective cross-sectional core area of the primary encircling coil (outside diameter of coil form, not inside diameter that is adjacent to the object). For internal probe electromagnetic testing, the ratio of the effective cross-sectional area of the primary internal probe coil to the cross-sectional area of the tube interior.

Fill factor effect: Effect of fill factor on coupling between coil and test object.

Flaw: Rejectable or unintentional anomaly. See also *Defect* and *Discontinuity.*

Flux density: See *Magnetic flux density.*

Flux leakage: See *Magnetic flux leakage field, Magnetic flux leakage technique*, and *Magnetic flux meter.*

Flux meter: See *Magnetic flux meter.*

Hall detector: Semiconductor element that produces an output electromotive force proportional to the product of the magnetic field intensity and a biasing current.

Hall effect: Potential difference developed across a conductor at right angles to the direction of both the magnetic field and the electric current. Produced when a current flows along a rectangular conductor subjected to a transverse magnetic field.

Hertz: Measurement unit of frequency, equivalent to one cycle per second.

Heterogeneity: The quality or state of being nonuniform or dissimilar.

Horseshoe coil: Probe coil in which the ferrite core of the coil is horseshoe shaped. Also called a *U-shaped coil.*

Hysteresis: Apparent lagging of the magnetic effect when the magnetizing force acting on a ferromagnetic body is changed; phenomenon exhibited by a magnetic system wherein its present state is influenced by its previous history.

Hysteresis loop: Curve showing flux density *B* plotted as a function of magnetizing force *H* as magnetizing force is increased to the saturation point in both negative and positive directions sequentially. The curve forms a characteristic shaped loop.

IACS: International Annealed Copper Standard.

Impedance: Opposition that a circuit presents to the flow of an alternating current, specifically the complex quotient of voltage divided by current.

Impedance analysis: In electromagnetic testing, an analytical technique that consists of correlating changes in the amplitude, phase, quadrature components or all of these of a complex test signal voltage to the condition of the test object.

Impedance-plane diagram: Graphical representation of the locus of points indicating the variations in the impedance of a test coil as a function of a parameter, such as conductivity or liftoff.

Indication: Nondestructive test equipment response to a discontinuity that requires interpretation to determine its relevance. Compare *Defect, Discontinuity,* and *False indication.*

Indication, discontinuity: Visible evidence of a material discontinuity. Subsequent interpretation is required to determine the significance of an indication.

Indication, nonrelevant: Indication due to misapplied or improper testing. May also be an indication caused by an actual discontinuity that does not affect the usability of the test object (a change of section, for instance).

Indication, relevant: Indication from a discontinuity (as opposed to a nonrelevant indication) requiring evaluation by a qualified inspector, typically with reference to an acceptance standard, by virtue of the discontinuity's size, shape, orientation, or location.

Induced current technique: See *Current induction technique.*

Inductor: Device consisting of one or more associated windings, with or without a magnetic core, which impedes the flow of current.

Initial permeability: Slope of the induction curve at zero magnetizing force as the test specimen begins to be magnetized from a demagnetized condition (slope at the origin of the *B,H* curve before hysteresis is observed).

Inserted coil: See *Inside diameter coil.*

Inside coil: See *Inside diameter coil.*

Inside diameter coil: Coil or coil assembly used for electromagnetic testing by insertion into the test piece, as with an inside probe for tubing. Also called *inserted coil.*

International Annealed Copper Standard (IACS): Conductivity measurement system in which the conductivity of annealed, unalloyed copper is arbitrarily rated at 100% and in which the conductivities of other materials are expressed as percentages of this standard. See also *Conductivity* and *Percent International Annealed Copper Standard.*

Leakage flux: (1) Magnetic flux of the coil that does not link with the test object. (2) Magnetic flux that leaves a saturated or nearly saturated specimen at a discontinuity.

Level, acceptance: See *Acceptance level.*

Level, rejection: See *Rejection level.*

Liftoff: Distance between the probe coil and the test object.

Liftoff effect: In an electromagnetic test system output, the effect observed due to a change in coupling between a test object and a probe whenever the distance between them is varied.

Longitudinal magnetic field: Magnetic field wherein the flux lines traverse the component in a direction essentially parallel with its longitudinal axis.

Magnetic field: Distribution of a vector quantity that is a measure of an exerted magnetic force.

Magnetic field indicator: Device used to locate or determine relative intensity of a flux leakage field.

Magnetic field intensity: Strength of a magnetic field at a specific point. Measured in ampere per meter.

Magnetic flux density: Normal magnetic flux per unit area, measured in gauss (G) or tesla (T).

Magnetic flux leakage: Excursion of magnetic lines of force from the surface of a test object.

Magnetic flux leakage field: Magnetic field that leaves or enters the surface of an object.

Magnetic flux leakage testing (MFL): Nondestructive testing method for the detection and analysis of a surface discontinuity or near-surface discontinuity using the flux that leaves a magnetically saturated, or nearly saturated, test object at a discontinuity.

Magnetic flux meter: Electronic device for measuring the field strength of magnetic flux leakage.

Magnetic particle testing (MT): Nondestructive test method using magnetic leakage fields and indication materials to disclose surface and near-surface discontinuities.

Modulation: Process of imparting information to a carrier signal by the introduction of amplitude or phase perturbation.

Multifrequency: Two or more frequencies applied sequentially or simultaneously to the test coil.

Multifrequency technique: Use of the response of a test object to more than one frequency, usually to separate effects that would be indistinguishable at a single frequency.

NDT: Nondestructive testing.

Noise: In electromagnetic testing, any nonrelevant signal that tends to interfere with the normal reception or processing of a desired discontinuity signal. Such noise signals may be due to an extraneous source or generated by heterogeneities in the test part that are not detrimental to the use of the part.

Nondestructive testing (NDT): Determination of the physical condition of an object without affecting that object's ability to fulfill its intended function. Nondestructive test methods typically use an appropriate form of energy to determine material properties or to indicate the presence of material discontinuities (surface, internal, or concealed).

Nonferromagnetic material: Material not magnetizable and essentially not affected by magnetic fields.

Percent International Annealed Copper Standard (%IACS): Measurement of conductivity as a percentage of the conductivity of pure copper, arbitrarily rated at 100%. See also *International Annealed Copper Standard.*

Period: Absolute value of the minimum interval after which the same characteristics of a periodic waveform or a periodic feature repeat.

Permeability: Ratio of magnetic induction to magnetizing force. This relationship is either (1) absolute permeability, in general the quotient of magnetic induction divided by the magnetizing force, or (2) relative permeability (or specific permeability), a pure number that is the same in all unit systems. The value and dimension of absolute permeability depend on the system of units used. In anisotropic media, permeability is a matrix.

Phase analysis: Analytical technique that discriminates between variables in a part undergoing electromagnetic testing by the different phase angle and amplitude changes that these conditions produce in the test signal. See also *Phase detection.*

Phase angle: Angular equivalent of the time displacement between corresponding points on two sine waves of the same frequency.

Phase detection: Derivation of a signal whose amplitude is a function of the phase angle between two alternating currents, one of which is used as a reference.

Pulse technique: Multifrequency technique in which a broadband excitation such as an impulse is used. Either the frequency components are extracted and analyzed or the interpretation is based directly on characteristics of the time domain waveform.

Qualification: Process of demonstrating that an individual has the required amount and the required type of training, experience, knowledge, and capabilities.

Qualified: Having demonstrated the required amount and the required type of training, experience, knowledge, and abilities.

Recommended practice: Set of guidelines or recommendations.

Recommended Practice No. SNT-TC-1A: Set of guidelines published by the American Society for Nondestructive Testing, for employers to establish and conduct a nondestructive testing personnel qualification and certification program.

Reference standard: Reference used as a basis for comparison or standardization. In tube testing, a tube with artificial discontinuities used for establishing the test sensitivity setting and for periodically checking and adjusting the sensitivity setting as required. See also *Acceptance standard.*

Rejection level: Value established for a test signal above or below which test specimens are rejectable or otherwise distinguished from the remaining specimens. This level is different from the rejection level as defined for ultrasonic and other test systems.

Relative permeability: Ratio of the permeability of the material to the permeability of a vacuum.

Sensing coil: Coil that detects changes in the flow of eddy currents induced by an excitation coil; sensing and excitation coils can be one and the same. Also called *detector coil.*

Signal: Physical quantity, such as electrical voltage, that contains relevant information.

Signal-to-noise ratio: Ratio of signal values (responses that contain relevant information) to baseline noise values (responses that contain nonrelevant information). See *Noise.*

Skin depth: Standard depth of penetration. See also *Skin effect.*

Skin effect: Phenomenon wherein the depth of penetration of electrical currents into a conductor decreases as the frequency of the current is increased. At very high frequencies, the current flow is restricted to an extremely thin outer layer of the conductor. See *Standard depth of penetration.*

SNT-TC-1A: See *Recommended Practice No. SNT-TC-1A.*

Specification: Set of instructions or standards invoked by a specific customer to govern the results or performance of a specific set of tasks or products.

Standard: (1) Physical object with known material characteristics used as a basis for comparison or standardization; reference standard. (2) Concept established by authority, custom, or agreement to serve as a model or rule in the measurement of quantity or the establishment of a practice or procedure. (3) Document to control and govern practices in an industry or application, applied on a national or international basis and usually produced by consensus. See also *Acceptance standard and Reference standard.*

Standard depth of penetration: In electromagnetic testing, the depth at which the magnetic field intensity or intensity of induced eddy currents has decreased to 37% of its surface value. The square of the depth of penetration is inversely proportional to the frequency of the signal, the conductivity of the material, and the permeability of the material. See also *Skin effect.*

Standardization, instrument: Adjustment of an NDT instrument, using an appropriate reference standard, to obtain or establish a known and reproducible response. (This is usually done prior to testing, but can be carried out anytime there is concern about the test or instrument response. See also *Calibration, instrument.*

Test coil: Section of a coil assembly that excites or detects the magnetic field in the material under electromagnetic test.

Test frequency: In electromagnetic testing, the number of complete cycles per unit time of the alternating current applied to the primary test coil.

Three-way sort: Electromagnetic sort based on a test object signal response above or below two levels established by three or more calibration standards.

Threshold level: Setting of an instrument that causes it to register only those changes in response greater or less than a specified magnitude.

Through-transmission: Of or pertaining to electromagnetic techniques where the excitation field penetrates the test object so that the detected signal is responsive to features external to or on the opposite surface. The primary coil and secondary coil are positioned on opposite sides of the product to be inspected.

Volt (V): Measurement unit of electric potential.

Index

Figures and tables are denoted respectively by *f* and *t* after the page numbers. Terms that start with numerals are sorted as though spelled out.

B

C

D

F

J

L

M

Q

R